MANUAL OF
Electronystagmography

MANUAL OF
Electronystagmography

HUGH O. BARBER, M.D., F.R.C.S.(C)

Professor of Otolaryngology, University of Toronto;
Head of the Department of Otolaryngology and Director
of the Dizziness Unit, Sunnybrook Medical Centre,
Toronto, Ontario, Canada

CHARLES W. STOCKWELL, Ph.D.

Associate Professor of Otolaryngology,
The Ohio State University College of Medicine,
Columbus, Ohio

SECOND EDITION

with **343** illustrations

The C. V. Mosby Company

ST. LOUIS • TORONTO • LONDON   1980

SECOND EDITION

Copyright © 1980 by The C. V. Mosby Company

All rights reserved. No part of this book may be reproduced
in any manner without written permission of the publisher.

Previous edition copyrighted 1976

Printed in the United States of America

The C. V. Mosby Company
11830 Westline Industrial Drive, St. Louis, Missouri 63141

Library of Congress Cataloging in Publication Data

Barber, Hugh O
    Manual of electronystagmography.

    Bibliography: p.
    Includes index.
    1. Electronystagmography. I. Stockwell, Charles W.,
1940-   joint author. II. Title. [DNLM: 1. Electronystagmography.   WW143 B234m]
RE748.B35   1980   617.7′6207547   80-17349
ISBN 0-8016-0449-4

C/CB/B 9 8 7 6 5 4 3 2 1   01/C/049

To
Grace Wright
colleague, teacher, friend

Preface

The recording of eye movements, including nystagmus, has become a relatively routine practice in the clinical investigation of patients with vertigo and balance disturbance. Test equipment, procedures, and interpretation of tracings are not yet standardized, although efforts to these ends are in progress in the United States. The methods advocated in this book seem logical, scientifically defensible, and practical to us, and we hope that readers will adopt them.

Much new information on central nervous system mechanisms of orientation, such as those related to maintaining visual grasp of the environment during head and body movement, has been obtained in recent years. Chapter 2, which has been extensively rewritten, attempts to make this important material available to the clinical electronystagmographer. In Chapter 4 new material has been introduced regarding measurement of nystagmus intensity. Chapter 5 presents new information and reflects a number of deletions, replacements, and additions of ENG tracings. In Chapter 9 normal caloric values from different laboratories have been tabulated, our only example of caloric perversion is shown, and a discussion of the difficulties involved in interpretation of the vertical lead in caloric testing has been added. We hope that these modifications enhance the quality of the book and render it a more valuable reference.

Audiologists and technicians often make laboratory tests of vestibular and oculomotor function and interpret the ENG recording of the test results. Even experienced otoneurologists may have difficulty in making reliable diagnostic inferences from ENG tracings alone; there is certainly the risk of misleading a referring physician by overinterpretation of records, particularly if they are of marginal quality. The physician's responsibility is to inform himself or herself of the value of certain reliable ENG findings and to relate these findings to the other medical data available for diagnosis. We have tried to keep this dual responsibility in mind while writing this book.

Hugh O. Barber
Charles W. Stockwell

Contents

Chapter 1

Introduction

USES OF ELECTRONYSTAGMOGRAPHY (ENG)

The physician gains useful information about the patient who complains of disturbed equilibrium by observing the patient's eye movements during certain kinds of visual and vestibular stimulation. At times these observations provide the only physical findings that support the patient's complaint. They also help the physician define the anatomic location of the patient's disorder. By observing eye movements the physician is often able to distinguish between a peripheral vestibular disorder and one located within the central nervous system and is sometimes able to lateralize a peripheral disorder or to further localize a central nervous system disorder.

In the past, the physician examined eye movements merely by direct observation of the patient's eyes. Considerable information can be gained by this method, but important signs are sometimes missed because the physician cannot prevent the patient from fixating. Visual fixation has a powerful suppressive effect on some types of nystagmus, reducing the nystagmus by as much as 90 percent below its intensity when fixation is denied and thereby making it difficult to detect by direct observation. Moreover, certain types of brain lesions and certain drugs impair or abolish the visual suppression effect; this phenomenon cannot be appreciated unless nystagmus is observed both when visual fixation is allowed and when it is denied.

In an attempt to minimize visual fixation, some physicians examine eye movements while their patients wear Frenzel's glasses, which are 20-diopter lenses mounted together with two small lights in a gogglelike frame that fits snugly against the patient's head. Frenzel's glasses magnify and illuminate the patient's eyes and, if used in a darkened room, reduce his ability to fixate.

Unfortunately, Frenzel's glasses do not entirely abolish fixation; therefore, physicians have watched with interest the development of other methods of monitoring eye movements. A number of methods are available, but the one best suited to the needs of physicians is *electro-oculography*. This method provides a permanent record of eye movements, either with the eyes open or closed, in light or in darkness. Moreover, the tracings that it yields permit the physician to study eye movements at leisure, to make direct comparisons between successive tests, and to make quantitative measurements of nystagmus intensity.

1

Electro-oculography has long been widely used in research on eye movements. Physicians who use the method for clinical study of dizziness and balance disturbances prefer to call it by another name, electro*nystagmo*graphy (ENG), because they use it primarily to monitor *nystagmus*.

METHOD OF ENG

ENG owes its existence to the fact that the eye is a battery. The cornea is the positive pole, the retina is the negative pole, and the potential difference between the two poles is normally at least 1 mv. This electrical potential, called the *corneo-retinal potential,* creates in the front of the head an electrical field that changes its orientation as the eyeballs rotate. These electrical changes can be detected by electrodes placed on the skin. When the changes are amplified and used to drive a writing instrument, a tracing of eye position is obtained.

Electrodes can be arranged on the skin in a number of ways, but the scheme shown in Fig. 1-1 is standard for clinical purposes. Horizontal eye position is

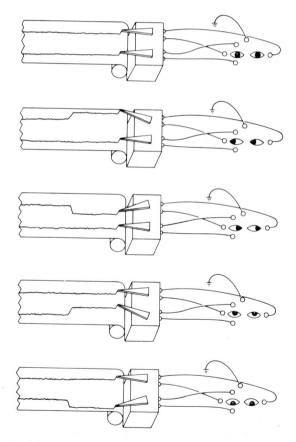

Fig. 1-1. The clinical ENG recording arrangement and the effects of eye movement.

monitored by two electrodes placed bitemporally, that is, one on the right temple and the other on the left temple. When the eyes are at midposition, there is a certain voltage between the electrodes that serves as a baseline. The recording system is arranged so that as the eyeballs move to the right, the change in voltage between electrodes causes an upward pen deflection, and as the eyeballs move to the left, the change in voltage causes a downward deflection. A second pair of electrodes, one above and the other below one of the eyes, is used to monitor vertical position of the eyes on another channel of the recording instrument. Upward movement produces an upward pen deflection, and downward movement produces a downward pen deflection. An additional electrode, usually on the forehead, is connected to the chassis of the recording instrument and serves as a ground.

This electrode arrangement faithfully monitors only conjugate eye movements. The bitemporal pair of electrodes records from both eyes simultaneously; therefore, it does not discriminate disconjugate movements. The vertical pair records from only one eye, and the assumption is made that the movements of the opposite eye are the same. One may detect disconjugate eye movements by recording from each eye on a separate recording channel.

The ENG method cannot detect torsional movements (rotations of the eye about the visual axis), because such movements produce no change in the orientation of the corneoretinal potential. Some torsional movements are clinically significant; therefore, failure to detect torsional movements must be regarded as a limitation of the ENG method. However, this limitation is not a severe one, because nearly all clinically significant torsional movements also have horizontal or vertical components that can be detected in the ENG tracing.

SEVEN BASIC TESTS

The ENG method is used today to monitor eye movements of the patient as he undergoes a battery of vestibular and oculomotor tests. In fact, the method has become so firmly identified with this test battery that the term *ENG* is commonly used to refer to either the method or the test battery or both. There is no universal agreement about which particular tests should comprise the battery, but seven tests are almost always included. These seven tests are briefly described here; they are considered in detail in Chapters 5 through 9.

1. *Gaze test.* Eye movements are recorded as the patient looks straight ahead, looks to the right, looks to the left, looks up, and looks down, with the eyes both open and closed. The tracing is inspected for the presence of nystagmus under any of these conditions.

2. *Saccade test.* Two pairs of dots are placed on the wall. One pair is placed so that its members are separated by a known distance (usually 20° visual angle) and so that an imaginary line between them is horizontal. The second pair is placed so that its members are separated by the same distance and so that an imaginary line

between them is vertical. The patient's eye movements are recorded as he looks back and forth between the two members of the horizontal pair and then back and forth between the two members of the vertical pair. The primary purpose of this procedure is to calibrate the recording system, but the tracing is also inspected for defects of saccadic eye movements.

3. *Tracking test.* Eye movements are recorded while the patient follows a slowly moving visual target. The tracing is inspected for defects of pursuit eye movement.

4. *Optokinetic test.* Eye movements are recorded while the patient watches vertical stripes moving horizontally at several different speeds to the right and then to the left. The tracing is inspected to determine whether the nystagmus generated by this stimulus becomes stronger as stimulus speed increases and whether it is stronger in one direction than it is in the other.

5. *Positional test.* Eye movements are recorded with the patient's eyes both open and closed, after the patient has been placed in various positions (usually sitting, supine, right lateral, left lateral, and head-hanging). The tracing is inspected for the presence of nystagmus in any position.

6. *Hallpike maneuver.* The patient is moved rapidly from sitting to the head-hanging right lateral position, then returned to sitting, then moved rapidly to the head-hanging left lateral position, and again returned to sitting. The tracing is inspected for nystagmus following each movement.

7. *Caloric test.* Each ear is irrigated twice, once with water (or air) that is above body temperature and once with water (or air) that is below body temperature. Each irrigation affects the vestibular receptors of the irrigated ear and provokes a horizontal nystagmus response. The responses provoked by right ear irrigation are compared with those provoked by left ear irrigation to determine whether the sensitivities of the left and right vestibular mechanisms are equal. Other features of the responses are noted as well.

These tests are not new. In one form or another, all have been a part of the otoneurologic examination for many years, although formerly the physician had monitored ocular responses by visual observation of the patient's eyes and thus had been forced to accept the limitations imposed by this method. Now that ENG is used to monitor ocular responses, the diagnostic power of these tests is substantially increased: physicians can identify significant pathologic nystagmus that they could not detect before, they can quantify it, they can make comparisons between successive tests, and they can evaluate the effects of visual fixation.

Patients who complain of dizziness commonly undergo ENG as part of their clinical evaluation, and some patients who are not dizzy need ENG. Before referring a patient for ENG, the physician should have formulated the question about the patient that ENG may be able to answer. Like other diagnostic tests, ENG is designed to answer three questions:

1. *Does a lesion exist?* ENG is often successful in answering this question,

provided that the physician is inquiring specifically about a vestibular or oculo-motor lesion. Positive findings may yield the only evidence that such a lesion exists. Negative findings, on the other hand, generally cannot be used to exclude the presence of a lesion.

2. *If a lesion exists, what is its location?* ENG is sometimes successful in answering this question. Although ENG findings are often nonlocalizing, some abnormalities permit a discrimination to be made between a peripheral lesion (labyrinth or eighth nerve) and a central nervous system (CNS) lesion. Certain abnormalities permit lateralization of a peripheral lesion, but none discriminates a labyrinthine lesion from one in the eighth nerve. Certain other abnormalities permit localization within the CNS.

3. *If a lesion exists, what is its cause?* ENG is rarely successful in answering this question.

By themselves the results of these tests almost never provide a definitive diagnosis. To be sure, strong relationships, both positive and negative, exist between certain test results and causes. Nevertheless, the results of these tests virtually always must be interpreted by the referring physician in light of the patient's history and other findings in order to obtain a diagnosis.

ROLES OF PHYSICIAN AND TECHNICIAN

It is impractical and unnecessary for physicians to conduct the ENG tests; frequently a technician can perform this task. The primary responsibility of the technician is to produce an interpretable ENG tracing. In addition, the technician usually calculates caloric test results and makes notes of observations and events that could influence test interpretation. Beyond that, his duties depend on the expectations of his employers. One technician may work for an otoneurologist who wishes to supervise the testing closely and to interpret the tracings himself. Another may work for a physician whose major interest lies outside otoneurology. The technician who works for the latter type of physician may be expected to interpret the tracing and issue a report to his employer or even directly to the referring physician. In such a report, the technician would describe the findings and any localizing value but would not offer suggestions regarding etiologic diagnosis. Diagnosis is the physician's responsibility.

The role of the ENG technician is similar in many respects to that of the medical audiologist. Both must be able to enlist the cooperation of patients and to make precise measurements using sophisticated electronic devices. In other respects, however, the roles of the ENG technician and the medical audiologist differ. One difference is that the audiologist works in a well-established specialty with widely recognized training standards. By comparison, ENG is a recent innovation. A few training programs for technicians are available, but training standards do not yet exist. Another difference is that the audiologist provides rehabilitation services to self-referred patients, whereas the ENG technician does not accept self-referrals.

The results of ENG tests are meaningful only after consideration by a physician in relation to the patient's history and other test results. Moreover, parts of the ENG procedure may be hazardous for certain patients, and this determination is one for which a physician must be responsible. In general, the ENG technician works under closer supervision of a physician than the audiologist does, although the well-trained and experienced technician may operate with some degree of autonomy.

Chapter 2

Applied neurophysiology

CONTROL OF EYE MOVEMENTS

Electronystagmography (ENG) is the study of eye movements. This chapter describes how eye movements are controlled and explains how certain neural lesions cause them to be deranged.

Most eye movements are complex, but they are made up of identifiable components. Some of these components are studied by users of ENG and others are not. In order to specify which components are studied, it is useful to introduce a classification system. Most physiologists accept the scheme shown in Table 2-1, which classifies eye movement components according to *amplitude* and *conjugation*.

An eye movement may have a *macromovement* or a *micromovement* component (or both). A macromovement has an amplitude larger than 1° of displacement, and a micromovement has an amplitude smaller than 1° of displacement.

An eye movement may also have a *version* or a *vergence* component (or both). A pure version occurs when the gaze is shifted from one visual target to another at the same distance from the observer. It is a conjugate movement; that is, the eyes move exactly together so that the angle between the two lines of sight remains the same (Fig. 2-1, *A*). Large-amplitude versions are called *macroversions*, and small ones, *microversions*. A pure vergence is performed when the gaze is shifted from one target to another that is either nearer to or farther from the observer. It is a disconjugate movement; that is, the angle between the two lines of sight changes during the movement in order to keep the retinal images of the target centered on the two foveae (Fig. 2-1, *B*). Large vergences are called *macrovergences*, and small ones, *microvergences*. Most eye movements are neither pure versions nor pure vergences but a combination of the two (Fig. 2-1, *C*).

Table 2-1. Classification of eye movement components

| Amplitude | Conjugation | |
	Version	*Vergence*
Macromovement	Macroversion	Macrovergence
Micromovement	Microversion	Microvergence

7

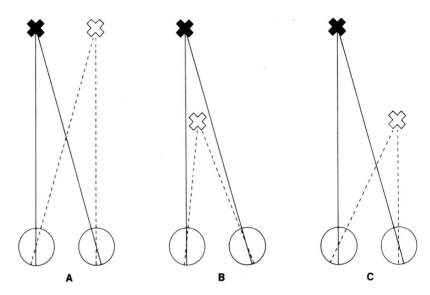

Fig. 2-1. **A,** Pure version. **B,** Pure vergence. **C,** Both version and vergence.

ENG is routinely used to monitor only *macroversions*. Microversion components and microvergence components cannot be monitored, because they are too small to be detected by the ENG method. Macrovergence components can be detected when the electrodes are placed to record separately from each eye, but they are not usually monitored for clinical purposes. When separate recordings are made from each eye, the purpose is virtually always to detect disconjugate movements arising from a pathologic condition of the version system. Because the ENG method is used to evaluate macroversion components, only this class of eye movement is considered here. Furthermore, the discussion will deal primarily with *horizontal* macroversion components. Vertical macroversion components will be considered, but only in passing. The reason for this emphasis on horizontal macroversions is that the neural mechanisms of this component have been worked out in great detail, whereas the mechanisms of the vertical macroversion component are almost entirely unknown.

ORGANIZATION OF NEURAL MECHANISMS

The neural mechanisms that control horizontal macroversions can be represented as a series of discrete elements (Fig. 2-2). The *motor system* moves the eyeballs. It is controlled by signals from the premotor system. The motor and premotor systems together are designated the *final common pathway.*

The final common pathway receives commands from four control systems — *vestibular, optokinetic, pursuit,* and *saccade* — and each control system performs a specific function. To understand the functions of these four control systems, one

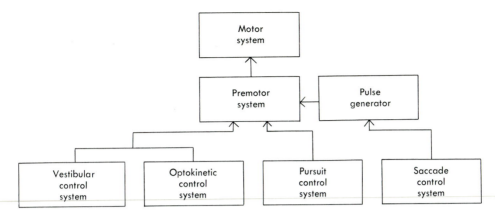

Fig. 2-2. Components of the control system for macroversions.

must first consider why we need eye movements in the first place. We need them because the resolving power of most of the retina is poor. Spatial sensitivity is good only in the fovea, a small patch of densely packed cones located at the center of each retina. To see a visual target in fine detail, we must aim our eyes so that the images of the target fall exactly on the two sensitive foveae. Two of the control systems, vestibular and optokinetic, keep visual targets on the foveae *despite movements of the head.* Each of these two systems makes its own contribution, and together they generate compensatory eye movements that keep the eyes aimed at the target. The third control system, pursuit, keeps visual targets on the foveae *despite movements of the target.* It monitors target motion and generates tracking motions that keep the eyes locked onto it. The fourth control system, saccade, places *new targets* on the foveae. Whenever an interesting target appears in the periphery of the visual field, it fires the *pulse generator,* which generates a rapid eye movement that places the target on the fovea.

In the remaining sections of this chapter, the salient features of these neural elements are described.

FINAL COMMON PATHWAY
Motor system

The motor system is composed of the extraocular muscles and the oculomotor nerves.

Extraocular muscles. Each eye is moved in its orbit by six muscles: four *rectus muscles* (medial, lateral, superior, and inferior) and two *oblique muscles* (superior and inferior) (Fig. 2-3). These six muscles form three pairs, the two members of a given pair exerting their actions in approximately the same plane but in opposite directions. The medial rectus and lateral rectus muscles form one pair, lying in the horizontal plane when the eye is in the straight ahead, or center gaze, position.

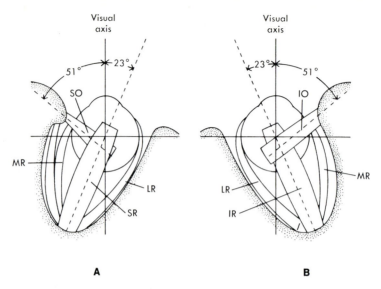

Fig. 2-3. The extraocular muscles of the right eye. **A,** Top view. **B,** Bottom view. *MR,* medial rectus; *LR,* lateral rectus; *SR,* superior rectus; *IR,* inferior rectus; *SO,* superior oblique; *IO,* inferior oblique. (Adapted from Burde, R. M.: The extraocular muscles: Part 1. Anatomy, physiology, and pharmacology. In Moses, R. A., editor: Adler's physiology of the eye, ed. 6, St. Louis, 1975, The C. V. Mosby Co., p. 94.)

The superior rectus and inferior rectus muscles, which form the second pair, lie in the vertical plane. When the eye is in the center gaze position, their plane of action makes an angle of 23° with the visual axis. The superior and inferior oblique muscles, which form the third pair, also lie in the vertical plane; when the eye is in the center gaze position, their plane of action makes an angle of 51° with the visual axis.

To move the eye, the agonist muscles (those used to pull the eye in the direction of movement) contract, and the antagonist muscles (those that pull the eye in the opposite direction) simultaneously relax. Precise specification of the direction of eyeball rotation produced by the contraction of a particular muscle is too complicated to be discussed in detail here. Most muscles induce a complex movement of the eyeball, and the direction of pull of a particular muscle changes from instant to instant as the position of the eyeball changes. Moreover, all eye muscles participate simultaneously in all eye movements, each muscle contributing its share to the movement as it progresses. For detail, the reader may consult the comprehensive treatment of this topic provided by Burde.[1]

It is, nevertheless, useful to consider specific eye movements that primarily depend on only one muscle. There are six such movements (Fig. 2-4). In each case, movement of the eye from the center gaze position to the position shown is produced chiefly by contraction of the indicated agonist muscle, although other

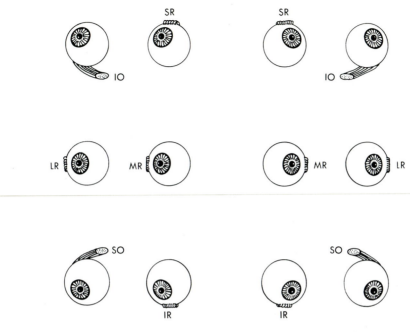

Fig. 2-4. The diagnostic gaze positions and their chief agonists. (Abbreviations are the same as in Fig. 2-3.)

muscles also participate to varying degrees. The corresponding muscles of the two eyes that contract when performing a given movement are called *yoke muscles*. For instance, when the eyes turn to the right, the right lateral rectus and left medial rectus muscles behave as if they were yoked together as a team, both pulling the eyes to the right. The six eye positions shown in Fig. 2-4 are called the *diagnostic gaze positions*. If one of the eyes cannot reach a particular gaze position, then the chief agonist muscle or its nerve supply is probably damaged.

Oculomotor nerves. The cell bodies of the nerves that supply the ocular muscles lie in the three oculomotor nuclei: *the third nucleus, the fourth (trochlear) nucleus,* and *the sixth (abducens) nucleus* (Fig. 2-5).[2] Each nucleus consists of a pair of cell masses, one member of the pair lying on each side of the midline of the brain. The pair of cell masses that comprises the third nucleus lies below the floor of the third ventricle and extends caudally almost to the trochlear nucleus. Nerve fibers from each side supply the ipsilateral medial rectus, inferior rectus, inferior oblique, and the contralateral superior rectus muscles. (Nerves from each side also innervate the levator, iris sphincter, and ciliary muscles of both eyes.) The fourth nucleus lies lateral and ventral to the aqueduct of Sylvius. Each side of this nucleus supplies the contralateral superior oblique muscle. The sixth nucleus is the most caudal of the three nuclei. It lies in the tegmental portion of the pons and each side supplies the ipsilateral lateral rectus muscle.

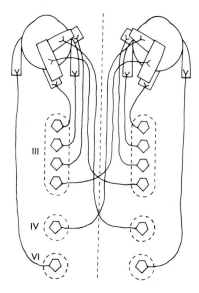

Fig. 2-5. Schematic diagram of the motor system. *III*, third nucleus; *IV*, fourth nucleus; *VI*, sixth nucleus.

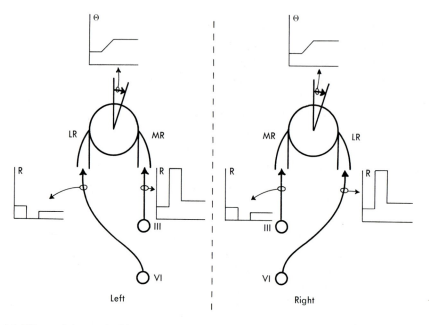

Fig. 2-6. Effects of changes in firing rate, *R*; of oculomotor neurons on eye position, *θ*. *LR*, lateral rectus muscle; *MR*, medial rectus muscle; *III*, third (oculomotor) nucleus; *VI*, sixth (abducens) nucleus.

Neurons in the oculomotor nuclei send to the ocular muscles a continuous barrage of neural impulses that keep the muscles in a steady state of partial contraction and thus maintain steady tension on the eyeball, even when it is motionless. The degree of contraction of a particular muscle depends on the amount of neural discharge it receives from the nerves that supply it.[3,4]

Fig. 2-6 shows how the firing rates of oculomotor neurons change when the eyes move from one position to another. In this example, the eyes make a sudden jump from center gaze to a new position 20° to the right. The chief agonists are the right lateral rectus muscle, which is supplied by neurons arising from the right sixth nucleus, and the left medial rectus muscle, which is supplied by neurons in the left third nucleus. During the movement, typical neurons in these nuclei exhibit a burst of activity, called a *pulse,* and then, after the movement is over, their firing rates settle back to a new, higher level, called a *step*. The shape of the total firing rate waveform is called a *pulse-step*.[3,4] The pulse overcomes the viscous drag of the eyeball and moves it rapidly to its new position; the step holds it there.

Not all neurons display exactly this firing rate waveform during the movement: some show only the pulse, others show only the step, and others show various combinations of the two.[5] In any case, the *average* of the firing rates of all agonist neurons is the pulse-step shown in Fig. 2-6.

The chief antagonists to this movement are the right medial rectus muscle and the left lateral rectus muscle, which relax. To produce the relaxation, the typical third nucleus neuron that supplies the right medial rectus muscle and the typical sixth nucleus neuron that supplies the left lateral rectus muscle cease firing altogether during the movement. Then, after the movement is over, they resume firing at a new, lower rate to help hold the eyeball in its new position. The shape of their firing-rate waveform is called an *inverse pulse-step*. Note that the inverse pulse-step is the mirror image of the pulse-step, except that the pulse is smaller. Agonist neurons fire at extremely high rates during the movement, whereas antagonist neurons merely stop firing; their firing rates cannot go below zero.

Effect of lesions in the motor system. As a general rule, lesions in the motor system cause *disconjugate* eye movements.[6] The misalignment of the visual axes is greatest when the eyes attempt to gaze in the direction that requires contraction of the paralyzed muscle (see Fig. 2-4), but it may also be present to a lesser degree in other gaze positions. As an example, Fig. 2-7 shows the effects of a lesion of the right sixth nerve. When the left eye is at center gaze (Fig. 2-7, *A*), the right eye is adducted because the contraction of the right medial rectus is unopposed by contraction of the paralyzed right lateral rectus. When right gaze is attempted (Fig. 2-7, *B*), the misalignment of the two visual axes is even greater, even though the right medial rectus is relaxes, because the paralyzed right lateral rectus is unable to contract.

The user of ENG is rarely called upon to evaluate disorders of the motor system itself. His concern is rather with premotor and control system disorders. However, he depends upon an intact motor system in order to make his evalua-

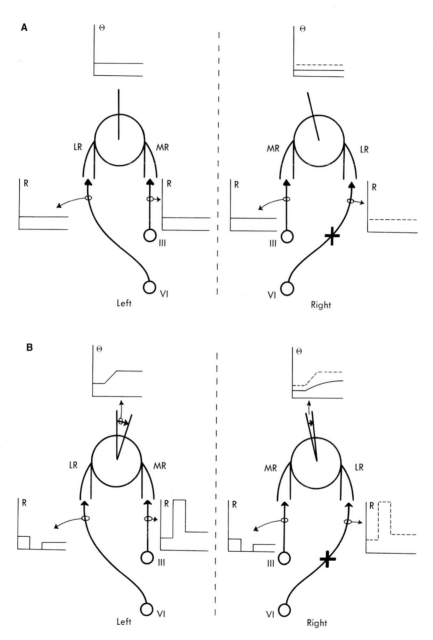

Fig. 2-7. Effects of a right sixth nerve lesion on eye position. **A,** Center gaze. **B,** Rightward gaze. (Abbreviations are the same as in Fig. 2-6) In the eye position, θ, and firing rate, R, graphs and dashed lines indicate desired values, solid lines indicate actual values.

tions. If the motor system is damaged, he must take this fact into account and perhaps modify his test procedure or interpretation accordingly. The lesion illustrated in Fig. 2-7 is easy to evaluate and, in such a case, the electronystagmographer may wish to make ENG recordings of only the left eye, which has an intact motor system. Other lesions may have more complicated effects, making it difficult to determine which muscle or even which eye is involved. Furthermore, the observed abnormality may not be caused by a motor system lesion at all but rather by a mechanical restriction of the eyeball in the orbit. In puzzling cases, a meaningful ENG examination may have to be deferred until a neuro-ophthalmologic evaluation is done, or it may simply be impossible.

Premotor system

The premotor system accepts commands from the eye movement control systems and distributes the signals required to execute these commands to the motor system. It lies in the *paramedian pontine reticular formation* (PPRF), immediately lateral to the sixth nucleus. The PPRF on the right side controls rightward movements and the PPRF on the left controls leftward movements.

The way in which the PPRF transforms commands into control signals is incompletely known at present, but the way it is presumed to work is shown in Fig. 2-8; this scheme is based on the one proposed by Robinson.[7,8] Admittedly it is speculative in some details, but it is logical and consistent with extant neurophysiologic evidence. Only the parts of the system that subserve horizontal movements are shown; vertical eye movement commands probably are processed in a similar manner, but our knowledge about them is scanty.

The eye movement shown in Fig. 2-8 is a sudden jump from center gaze to a new position 20° to the right, the same one shown in Fig. 2-6. We know that this movement is produced by a pulse-step to the agonist motoneurons and an inverse pulse-step to the antagonist motoneurons.[4] We also know that the command to the premotor system for this movement is a pulse.[9] The right PPRF receives the pulse and delivers it unaltered via the *straight-through pathway* to neurons in the medial longitudinal fasciculus (MLF). Excitatory MLF neurons facilitate neurons in the ipsilateral sixth nucleus neurons and contralateral third nucleus neurons. Inhibitory MLF neurons inhibit neurons in the contralateral sixth nucleus and ipsilateral third nucleus. (Inhibitory neurons are shown ascending in the MLF, but neurophysiologic evidence on this point is at present contradictory.[8] It is possible that inhibitory neurons ascend in pathways other than the MLF.) The left PPRF receives an inverse pulse and delivers it to excitatory MLF neurons, which disfacilitate neurons in the ipsilateral sixth nucleus and contralateral third nucleus. It also delivers the inverse pulse to inhibitory MLF neurons that disinhibit neurons in the contralateral sixth nucleus and ipsilateral third nucleus. These pathways produce the pulse component of the control signals.

The command signals also feed via a parallel pathway into neural *integrators* in the PPRF, which perform mathematical integration upon the command signals.

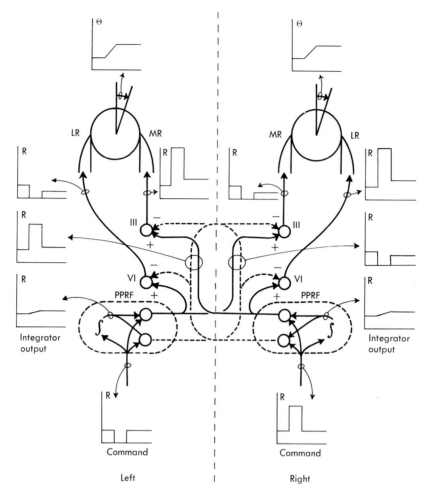

Fig. 2-8. Schematic diagram of the premotor system and effects of changes in firing rate, *R*, of premotor neurons on eye position, *θ*. *LR*, lateral rectus muscle; *MR*, medial rectus muscle; *III*, third (oculomotor) nucleus; *VI*, sixth (abducens) nucleus; *MLF*, medial longitudinal fasciculus; *PPRF*, paramedian pontine reticular formation.

However, the PPRF integrators are imperfect—they leak. By themselves, they are unable to provide a proper step (Fig. 2-9, *A*). Recent evidence[11, 12] suggests that the PPRF integrators are assisted in this task by additional neural circuits located in the cerebellum. How these circuits work is unknown, but they provide the extra integrating action needed to produce a near-perfect step (Fig. 2-9, *B*). The step is fed to the same MLF neurons that carry the pulse. The sum of the two signals—the pulse and the step—is the pulse-step signal that the MLF neurons carry to motoneurons.

As a result of this signal processing and distribution in the premotor system,

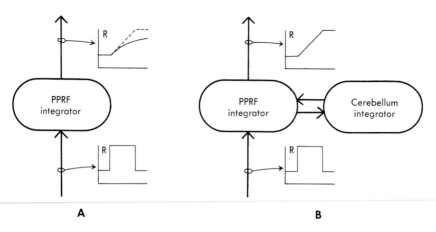

A   B

Fig. 2-9. A, "Leaky" integration of the paramedian pontine reticular formation *(PPRF)* integrator. Dashed line indicates desired output, solid line indicates actual output. **B,** Perfect integration of *PPRF* integrator when assisted by cerebellar integrator.

the agonist motoneurons (arising from right sixth and left third nuclei) receive a pulse-step of facilitation from the right PPRF and a pulse-step of disinhibition from the left PPRF, while the antagonist motoneurons (arising from the left sixth and right third nuclei) receive a pulse-step of inhibition from the right PPRF and a pulse-step of a disfacilitation from the left PPRF. The sum of all these signals produces the rapid eye movement to the right.

Effect of lesions in the MLF. A lesion in the medial longitudinal fasciculus between the third and sixth nuclei produces a highly distinctive defect known as internuclear ophthalmoplegia. The most salient features of this defect are a lag of the adducting eye and nystagmus in the abducting eye during rapid horizontal movements (see Figs. 5-24 and 6-6). The reason that this lesion causes such a defect is shown in Fig. 2-10.[8] As nerve fibers ascend in the MLF, they first send branches to the sixth nucleus, which innervates the lateral rectus muscles, then send branches to portions of the third nucleus that innervate the medial rectus muscles. In Fig. 2-10 the lesion has occurred on the left side. Thus, when the control signals are sent for a rapid eye movement to the right, they proceed normally to the sixth nuclei but are blocked on their journey to the third nuclei. As a result, the left medial rectus neurons fail to receive facilitation and the right medial rectus neurons fail to receive inhibition. The left eye therefore moves sluggishly and falls short of the target because the pulse to its medial rectus is much too small, but it continues to drift toward the target under control of the step. The right eye initially hits the target because the pulses to its muscles are the right size, but it immediately begins to slip back because the step to its medial rectus is too small to hold it there. The visual axis is now misaligned and off target, which triggers another eye movement that brings the eyes closer. This process is repeated until both eyes

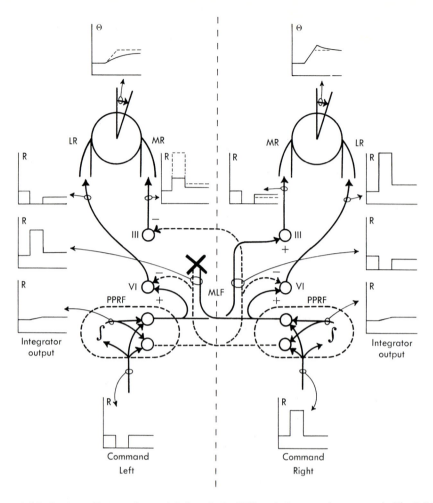

Fig. 2-10. Mechanism of internuclear ophthalmoplegia. (Abbreviations are the same as in Fig. 2-8.) In the eye position, θ, and firing rate, R, graphs and dashed lines indicate desired values, solid lines indicate actual values.

are on target. In the adducting eye, the weak pulses make the repeated attempts to hit the target appear to be one long, slow movement, whereas in the abducting eye, the series of jerks and drifts in the opposite direction give the appearance of nystagmus.

This explanation of internuclear ophthalmoplegia is somewhat speculative, and it has some weaknesses. For example, it predicts that nystagmus will appear in *both* eyes when they gaze ipsilateral to the lesion (to the left for the lesion shown in Fig. 2-10), since the steps to both third nuclei would be too small to hold the eyes on target.[8] This phenomenon has never been reported, but Robinson has suggested that it may be quickly compensated.

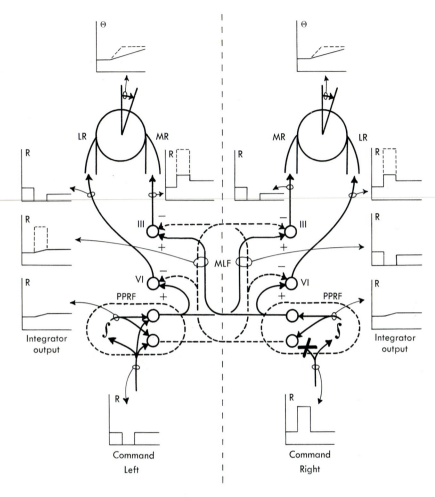

Fig. 2-11. Mechanism of saccadic slowing. (Abbreviations are the same as in Fig. 2-8.) In the eye position, θ, and firing rate, R, graphs and dashed lines indicate desired values, solid lines indicate actual values.

When the MLF lesion is bilateral, the defect is symmetric for both directions of eye movement and it is more severe, since all control signals from the PPRF to the third nuclei are blocked.

Effects of lesions in the PPRF. The effects of a PPRF lesion depend upon which pathways in the PPRF are interrupted. There are many possible combinations of lesions; only two will be described here — a lesion in the straight-through pathway and a lesion in the integrator.

A discrete lesion in the straight-through pathway of the right PPRF is illustrated in Fig. 2-11. Since the straight-through pathway is blocked, the only output from the right PPRF is the step from the integrator. Thus, for a quick eye movement to the right, the agonists receive pulseless facilitation plus normal disinhibition (from the left PPRF) and the antagonists receive pulseless inhibition plus normal disfacilitation. As a result, the steps are the proper size but the pulses are too small. Both eyes move sluggishly to their goal, producing a defect called *saccadic slowing*. Contralateral saccades (on the left in Fig. 2-11) are normal.

Lesions in the saccadic control system also produce saccadic slowing (see p. 118).

A lesion in the neural integrator of the right PPRF is illustrated in Fig. 2-12. Since the integrator does not integrate, the only output of the right PPRF is the pulse. Thus, for an eye movement to the right the agonists receive stepless facili-

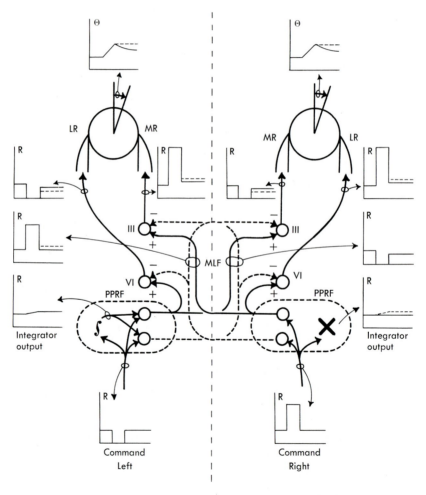

Fig. 2-12. Mechanism of gaze nystagmus. (Abbreviations are the same as in Fig. 2-8.) In the eye position, θ, and firing rate, R, graphs and dashed lines indicate desired values, solid lines indicate actual values.

tation plus normal disinhibition (from the left PPRF) and the antagonists receive stepless inhibition plus normal disfacilitation. As a result, the pulses are the proper size, but the steps are too small to hold the eyes in the deviated position. The eyes reach the target, but they begin to drift back toward a too-small "hold" position that is determined by step size. This state of affairs produces another attempt to put the eyes on target, but again they drift back, and so forth. The result is *gaze nystagmus*.

This defect has several features. First, the waveform of the drift back toward the "hold" position is *exponential,* since it is determined by the viscous drag of the eyeball. This exponential waveform is pathognomonic of a lesion of the neural integrator.[10] Second, the greater the initial eye deviation, the more pronounced the gaze nystagmus will be. The reason is that the step is always too small by a constant percentage of eye deviation. If, say, the step is 50 percent too small, the "hold" position for a 40° eye deviation would be 20° too small, whereas the "hold" position for a 10° deviation would be only 5° too small. Since the eyes always begin to slip back toward the "hold" position at an exponential rate, they would slip faster for a large deviation than for a small one. This phenomenon is il-

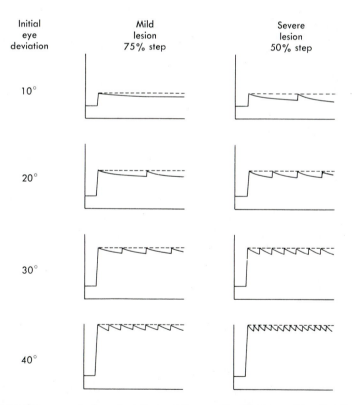

Fig. 2-13. Effects of the amount of eye deviation and the severity of lesion on intensity of gaze nystagmus. Dashed lines indicate desired eye position, solid lines indicate actual eye position.

lustrated in Fig. 2-13. Third, the more severe the lesion, the more pronounced the gaze nystagmus for a given eye deviation. This effect is also illustrated in Fig. 2-13. Two severities of lesion are shown. In the left column, a relatively mild lesion produces a step that is only 25 percent too small, and, in the right column a more severe lesion produces a step that is 50 percent too small. The more severe lesion causes gaze nystagmus to appear with smaller initial eye deviations and to be more pronounced at any given deviation. Fourth, the defect is always *bilateral.* For an eye movement contralateral to the lesion, the step would be too small by the same amount that it is for ipsilateral eye movement. In that case the agonist motoneurons would receive normal facilitation plus stepless disinhibition and the antagonist motoneurons would receive normal inhibition plus stepless disfacilitation.

Bilateral gaze nystagmus does not necessarily indicate a lesion in the PPRF itself. Recall that the PPRF integrator is inherently leaky (Fig. 2-9). It is assisted by another integrator located in the cerebellum.[11,12] Thus, gaze nystagmus could also point to a lesion in the cerebellum or its efferent pathways.

PULSE GENERATOR

The type of rapid eye movement we have been using as an example is called a saccade. It is the fastest eye movement of which the oculomotor system is capable. To produce this fast eye movement, the premotor system receives, as its command, a precisely timed pulse that rises almost instantaneously to a very high firing rate and shuts off exactly when the eyes reach the new target.[9] This signal is supplied by the pulse generator.

The pulse generator is probably located in the PPRF adjacent to or intertwined with the other circuits that make up the premotor system. It is not known exactly how it works, but a model proposed by Zee and Robinson[13] is plausible and is consistent with neurophysiologic evidence. According to the model, the pulse generator continuously receives a "desired eye position" signal. This signal is fed to *comparator neurons,* which compare it to "actual eye position" signals obtained from the output of the integrators. The firing rate of the comparator is proportional to the difference between the desired and actual eye position signals. The output of the comparator is fed to *burst neurons,* which have a very high gain, that is, they respond with very large changes in firing rate for small changes in firing rate of the comparator neurons that control them. Normally the burst neurons are inhibited by *pause neurons* to keep them from firing spuriously. To start the movement, the pause neurons are shut off by a "trigger" signal, allowing burst neurons to commence firing at a high rate and generate the pulse command to the premotor system. As the eyes reach the new target, the difference between the "desired eye position" signal and the integrator output signal goes to zero and the firing rate of the comparator returns to its baseline value. As the same time, the pause neurons are released from inhibition, suppressing the burst neurons and preventing them from oscillating.

This model of the burst generator has several attractive features: it makes use of cell types—burst neurons,[9] tonic (eye position) neurons,[14] and pause neurons[15] —that are known to exist. It also relieves the saccadic control system of the difficult task of preprogramming a pulse of just the right size to get the eyes from one visual target to another. Under this scheme, a saccade simply lasts until desired eye position and actual eye position coincide.

Effects of lesions in the pulse generator. The pulse circuit is inherently unstable because of its high gain. This instability is the price paid for the ability to move the eyes quickly. Because it is unstable, the pulse generator must be precisely tuned. There is a delay of approximately 10 msec in the eye position signal that feeds back from the integrator.[13] If the pause cells do not fire at an adequate level between saccades or if they do not immediately resume firing at the end of a saccade, this delay will cause the burst neurons to oscillate. Such oscillation causes an abnormality known as *ocular flutter,* in which the eyes oscillate briefly at the end of saccades or, occasionally, during steady fixation (see Fig. 6-10). Some individuals are able to voluntarily produce a type of ocular flutter called *voluntary nystagmus.*[16] Apparently they have learned to inhibit their pause cells at will.

VESTIBULAR AND OPTOKINETIC CONTROL SYSTEMS
Peripheral vestibular system

The vestibular control system monitors the position and movement of the head and sends this information to servomechanisms, which use it, along with information from other sensory systems, to control movements of the head, body, limbs, and eyes. The control of eye movements is perhaps not the most important function of the vestibular system, but it is certainly the one of greatest interest to users of ENG. The control signals sent by the vestibular system to the premotor system generate compensatory eye movements that help keep visual targets on the foveae during head movements.

The labyrinth. The vestibular receptors lie in the bony labyrinth, a complicated maze of interconnecting channels within the temporal bone. The bony labyrinth is filled with perilymph. Suspended within the bony labyrinth is the membranous labyrinth (Fig. 2-14), a sac of roughly the same shape as the bony labyrinth, filled with endolymph, a fluid with a chemical composition markedly different from that of perilymph. The membranous labyrinth is completely enclosed; nowhere does the endolymph mingle with the perilymph that surrounds it.

Vestibular sensory cells are embedded in the walls of the membranous labyrinth in five separate areas: in the cristae ampullares of the lateral, anterior vertical, and posterior vertical semicircular canals and in the macula utriculi and the macula sacculi.

Hair cells. Approximately 7,000 sensory cells are packed side by side in each of the cristae, 30,000 in the macula utriculi, and 16,000 in the macula sacculi.[17] These sensory cells are called hair cells, because cilia, or hairs, protrude from their surfaces (Fig. 2-15). Each hair cell possesses one long hair, called a kinocilium, and approximately 70 shorter, rodlike hairs, called stereocilia. The hairs

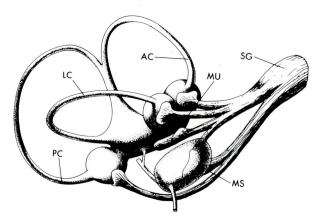

Fig. 2-14. The membranous labyrinth of the right ear. *LC*, lateral semicircular canal; *AC*, anterior vertical semicircular canal; *PC*, posterior vertical semicircular canal; *MU*, macula utriculi; *MS*, macula
sacculi; *SG*, Scarpa's ganglion.

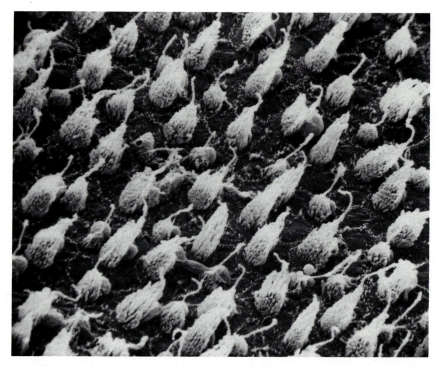

Fig. 2-15. Scanning electron photomicrograph of hair cells in the sensory surface of the macula utriculi.
Protruding from each hair cell is one long kinocilium and a bundle of shorter stereocilia. (× 1100.)
(Courtesy David Lim, M.D.)

project into the endolymphatic compartment of the labyrinth. Neighboring hair cells are separated by supporting cells, and tight junctions between hair cells and supporting cells prevent leakage of endolymph. The base of the hair cell, to which afferent and efferent nerve endings are opposed, hangs free in the perilymphatic compartment.

A comprehensive study of the morphology of the labyrinth was made by Lindeman.[18]

Hair cell function. Hair cells possess two important functional characteristics. First, most of them are spontaneously active in the absence of any external stimulation, continuously releasing neurochemical transmitter substance and provoking continuous activity in the apposed afferent nerves.[19] The total amount of spontaneous neural activity in the vestibular afferent nerve fibers is impressive. Approximately 20,000 afferent nerve fibers lie in each vestibular nerve, and the average spontaneous firing rate of each nerve fiber is approximately 75 impulses per second. Thus, the brain receives a tonic input on the order of 1.5 million impulses per sec from each labyrinth. The second important characteristic of hair cells is that they are directionally polarized.[20] When the hairs are bent in one direction, the hair cell is facilitated: its membrane is depolarized, less transmitter substance is released, and the firing rates of its afferent nerves are decreased. If the hairs are bent in the direction perpendicular to the plane of polarization, no change in the firing rate occurs.

The five vestibular receptors in each labyrinth can be classified according to the specific stimulus to which they respond. The semicircular canals form one type of receptor; they respond to angular accelerations of the head. The macula utriculi and macula sacculi, collectively known as the otolith organs, form the other type; they respond to linear accelerations of the head. This stimulus specificity is probably not absolute, however. Semicircular canals may in fact respond weakly to linear accelerations, and otolith organs may respond weakly to angular accelerations; however, it is doubtful that these nonspecific responses are functionally important.

Semicircular canals. Each semicircular canal is a continuous ring of fluid, which is blocked by the cupula, a gelatinous plug that completely fills the lumen of the canal in its ampullated portion (Fig. 2-16). Embedded in the cupula are the sensory hairs of the hair cells. When the head is at rest, the pressure of the fluid on both sides of the cupula is the same, and the cupula is at its neutral position. When the head undergoes angular acceleration, the rotation of fluid tends to lag behind the rotation of the canal walls. Because the fluid is blocked by the cupula, this lag creates a difference in pressure across the cupula, which makes it move, thus bending the sensory hairs embedded in it. As mentioned before, the effect of sensory hair bending on hair cell activity depends entirely on the directional polarization of the cells. In a semicircular canal receptor, all of the hair cells are polarized in the same direction.[18] Thus, when the cupula moves in one direction, all of the hair cells are excited; when it moves in the other direction, they are inhibited. The

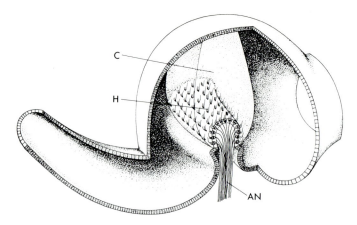

Fig. 2-16. Cross-sectional view of the ampulla of a semicircular canal. *C,* cupula; *H,* hair cells; *AN,* afferent nerves.

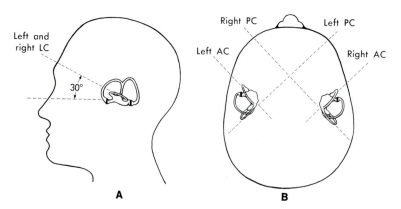

Fig. 2-17. Planes of the semicircular canals. **A,** Lateral canals. **B,** Anterior and posterior vertical canals. *LC,* lateral canal; *AC,* anterior vertical canal; *PC,* posterior vertical canal. The size of the canals is greatly exaggerated.

degree of hair cell excitation (or inhibition) is roughly related to the amount of cupula deflection, which in turn depends on the strength of the applied angular acceleration.

There are six semicircular canals, three in each temporal bone. These six canals function as three pairs lying in mutually perpendicular planes, with each pair maximally sensitive to angular accelerations in its own plane. The first pair consists of two lateral canals, one on the right and the other on the left. This pair lies in the horizontal plane when the head is inclined by 30° (Fig. 2-17, *A*). The hair cells in the two lateral canals are polarized in opposite directions, so that for any given head acceleration in the horizontal plane, the cells in one ear are facilitated and those in the other are inhibited. For example, if the head is accelerated to the right, both cupulae swing to the left, which causes the hair cells in the lateral canal of the right ear to be facilitated and those in the lateral canal of the left ear to be inhibited. The resulting asymmetry between the input of the two paired receptors is the signal to the brain that the head is being accelerated to the right. If the head accelerates to the left, the asymmetry is in the opposite direction: the cupulae swing to the right, causing the receptors in the lateral canal of the left ear to be facilitated and those in the lateral canal of the right ear to be inhibited.

Besides the lateral pair, there are two other semicircular canal pairs. One pair consists of the right anterior vertical canal and the left posterior vertical canal. The plane of this pair is shown in Fig. 2-17, B. The receptor polarization of the two members of this pair is also opposed. Forward angular acceleration of the head toward the right shoulder causes facilitation of the hair cells in the right anterior vertical canal and inhibition of those in the left posterior vertical canal. Backward acceleration over the left shoulder causes facilitation in the left posterior vertical canal and inhibition in the right anterior vertical canal. The third pair consists of the left anterior vertical canal and the right posterior vertical canal (Fig. 2-17, B). Receptor polarization is again opposed. Forward head acceleration causes left anterior facilitation and right posterior vertical inhibition. Opposite acceleration causes the opposite stimulation pattern. The general rule in remembering semicircular canal receptor response to angular acceleration is that the receptor in the leading ear is facilitated and the one in the trailing ear is inhibited.

Most natural head movements affect all the semicircular canals simultaneously. Since each canal pair responds only to the component of head acceleration that lies in its own plane, the brain can arrive at a correct interpretation of head angular motion by vector addition of the input from the three canal pairs.

Response to rotation. The semicircular canal system provides accurate information only for a narrow range of head movements, specifically, for those movements that occur in everyday life.[21] Whenever this system is driven by stimulations that exceed (either in magnitude or duration) those that occur during natural movements, it fails to provide correct information. Excessive stimulation might occur in amusement park rides or in aircraft or spacecraft maneuvers. To illustrate this point, one might consider a natural head movement, such as a 90° turn of the head to the right. This movement consists of a brief acceleration, followed immediately by a deceleration of equal magnitude and duration (Fig. 2-18, A). The initial acceleration to the right causes the cupulae of the two lateral canals to swing to the left, but they do not reach their fully deviated positions instantaneously. The inertia of the fluid and friction between the fluid and canal walls cause the response to be somewhat sluggish. By the time the cupulae have nearly reached their deviated positions, the deceleration begins, driving them back to their neutral positions again. The result of the sluggishness of the cupulae is that the receptor system performs a mathematical integration upon the acceleration stimulus and causes the response to resemble the velocity rather than the acceler-

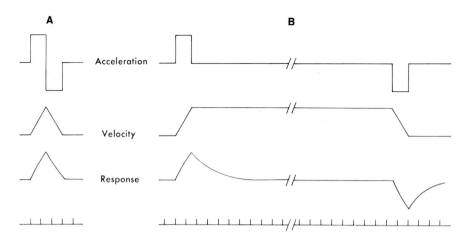

Fig. 2-18. **A,** Response of semicircular canals to brief rotation. **B,** Response to prolonged rotation.

ation of the head movement. The immediate sensation caused by vestibular stimulation is thus head velocity (not acceleration), and for natural head movements the information provided by the system is correct. It is not correct, however, for the stimulus shown in Fig. 2-18, *B*. This stimulus consists of the same initial angular acceleration, but this time the acceleration is followed by a period of constant speed rotation. After a few seconds of rotation, the person is decelerated to a stop. The vestibular sensation associated with this stimulus does not conform to angular velocity at all. The sensation is of increasing velocity as the cupulae are deviated during the acceleration; however, when acceleration ends, the cupulae return passively to their neutral positions and the sensation is of slowing down to a stop.

When the deceleration is applied, the sensation of turning in the opposite direction occurs. When the deceleration ceases, the cupulae swing back again to their neutral positions. The information that the vestibular system provides for this unnatural stimulus is completely erroneous. The consequence of incorrect information from the vestibular system is vertigo.

Response to caloric stimulation. The semicircular canal receptors are also sensitive to thermal gradients in the temporal bone.[22] Otoneurologists make use of this phenomenon to produce semicircular canal responses by irrigating the external auditory canal with warm or cool water (or air) in the caloric test (Chapter 8). The otoneurologist finds that caloric stimulation is more clinically useful than rotation, because the two labyrinths can be stimulated separately and their responses can be compared.

When used for clinical purposes, the caloric stimulus is usually administered while the patient is in the caloric test position, that is, supine with the head flexed by 30° (Fig. 2-19, *A*). When the head is in this position, the lateral semicircular canals are vertical and are thus maximally sensitive to thermal stimulation. The mechanism of stimulation is shown in Fig. 2-19, *B*. As the warm temperature

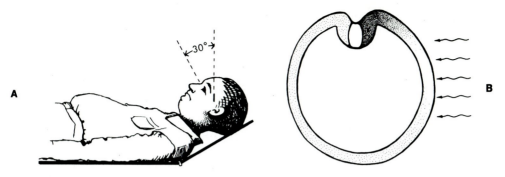

Fig. 2-19. A, Caloric test position. **B,** Effect of warm temperature irrigation on the right lateral canal.

wave traverses the temporal bone, the first site it reaches in the labyrinth is the most lateral portion of the lateral canal; thus it first warms the endolymph in this region. As the endolymph becomes warm, it becomes less dense and tends to rise, increasing the pressure on the lateral side of the cupula and making it deflect medially. Medial deflection of the lateral canal cupula facilitates its hair cells. For this reason, a warm temperature irrigation facilitates the irrigated ear.

A cool temperature irrigation produces a response in the opposite direction. The fluid in the most lateral portion is cooled, becomes more dense, sinks, and deflects the cupula laterally, thus inhibiting the hair cells.

Otolith organs. The otolith organs are linear accelerometers. They respond to linear accelerations because the sensory hairs of the hair cells are embedded in a gelatinous layer, which also contains a surface layer of calcite crystals called otoconia (Fig. 2-20). The otoconia are much heavier than the surrounding fluid; therefore, whenever the head undergoes a linear acceleration, they tend to lag behind, causing displacement of the gelatinous layer and bending of the hairs of the underlying hair cells. There are two otolithic organs in each labyrinth: one in the utricle, the macula utriculi; and the other in the saccule, the macula sacculi. Each receptor is maximally sensitive to the component of linear acceleration in the plane of its surface. The surface planes of these receptors are difficult to specify because they are bowl-shaped.[18] However, the "average" plane of the macula utriculi is roughly parallel to that of the lateral semicircular canal (horizontal when the head is tilted forward 30°). The "average" plane of the surface of the macula sacculi is approximately parallel to the sagittal plane of the head.

The hair cell polarization pattern of the otolith organs is much more complicated than that of the semicircular canals.[18] All hair cells of a given otolith receptor are not polarized in the same direction; instead, they are arranged so that they are oriented in all directions (Fig. 2-21). Thus, for a given linear acceleration in the plane of an otolithic receptor, some hair cells are facilitated, some are inhibited, and others are not affected at all. It is apparent that each direction of linear acceleration sets up a unique pattern of hair cell stimulation.

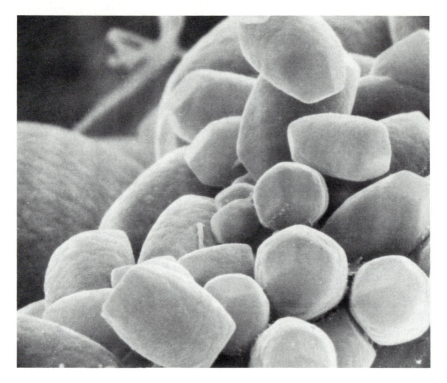

Fig. 2-20. Scanning electron photomicrograph of otoconia on the human macula utriculi. ($\times$ 4000.) (Courtesy David Lim, M.D.)

The way in which otolithic response patterns are analyzed by the central nervous system (CNS) is not known in detail. In this respect, our knowledge of otolith function lags far behind our knowledge of semicircular canal function. It is clear, however, that the major task of the otolith system is to sense the direction of gravity. Gravity is a linear acceleration like any other; it is a constant, upward-directed acceleration of approximately 32 ft/sec^2 (when one is on the surface of the earth). Information about the orientation of the hand relative to gravity provided by the otolith organs complements the information about angular acceleration of the head provided by the semicircular canals to give a complete picture of the orientation of the head in space.

The otolith organs are virtually ignored by otoneurologists. It is known that the otolith organs can initiate and influence eye movements, but there are no clinically useful ENG tests of otolith function. This lack of emphasis on otolith function in clinical otoneurology is based not on the conclusion that the otolith system is functionally unimportant but on a lack of knowledge concerning that function.

Central vestibular system

Anatomy. Afferent nerve fibers from the semicircular canals and otolith organs enter the central nervous system via the vestibular portion of the eighth nerve.

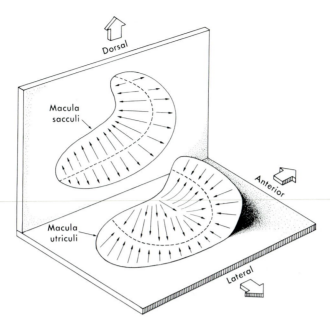

Fig. 2-21. Polarization of hair cells on the otolith organs. The direction of arrows indicates the direction of hair deflection that excites hair cells in that region of the receptor surface. (Data from Lindeman, H. H.: Adv. Anat. Embryol. Cell Biol. **42:**1, 1969.)

These nerve fibers are bipolar. Their cell bodies lie in Scarpa's ganglion; their peripheral processes synapse on the hair cells, and their central processes synapse on cells that lie in specific regions of the ipsilateral vestibular nuclei, although a few bypass the nuclei and proceed directly to the cerebellum.

The vestibular nuclei are complex aggregates of cell groups that lie in the floor of the fourth ventricle.[23] There is one group of nuclei on the right and one on the left side. At least ten distinct nuclei can be distinguished on each side, but four of these are most prominent: the superior, medial, lateral (or Deiter's), and descending nuclei. In addition to receiving afferent fibers from the labyrinth, these nuclei also receive input from other sources: from the spinal cord, the reticular formation, the contralateral vestibular nuclei, and especially the cerebellum. The influence of the cerebellum on the vestibular nuclei is primarily inhibitory.[24] In general, it may be said that the vestibular nuclei send efferent fibers to the same parts of the brain from which they receive afferent fibers.

Vestibulo-oculomotor pathways. The central vestibular pathways of greatest interest to users of ENG are the ones by which vestibular input influences eye movement. These pathways have been intensively studied in recent years. They are intricate but appear to be divided into two types: *direct pathways* from the vestibular nuclei to the oculomotor nuclei and *indirect pathways* from vestibular to oculomotor nuclei via the reticular formation and various other routes. An excellent summary of recent studies on vestibulo-oculomotor pathways is provided by Precht.[25]

There exists considerable controversy over the *functional* organization of these pathways. Robinson has offered a reasonable interpretation, which is presented, in highly simplified form, in Fig. 2-22. Although Robinson's interpretation is elegant and logical, the reader should be aware that it is only one of several possible views and that it rides roughshod over a great deal of anatomic and electrophysical evidence.

The head movement illustrated in Fig. 2-22 is a brief, quick movement to the left. The horizontal semicircular canal cupulae respond to this stimulus. The cupula on the left canal is deviated toward midline and thus its hair cells are facilitated; the cupula on the right is deviated away from midline and its hair cells are inhibited. As noted previously (p. 27), the semicircular canals produce a signal that is proportional to head velocity, at least for a quick head movement like the one illustrated here. This signal is relayed to the ipsilateral vestibular nucleus and then to the contralateral PPRF, where it passes through the two parallel pathways — straight-through and integrator.[7] The output of the PPRF is a *ramp-step,* which goes to the oculomotor neurons to produce an eye movement that keeps the eyes on target. The vestibulo-oculomotor system is beautifully designed to provide compensatory eye movements in response to a brief head movement. However, when the head movement persists beyond a second or two, the vestibular system is inadequate to the task. In that case, additional neural elements enter the picture.

The first thing that happens is illustrated in Fig. 2-23. The head rotation is again to the left, but it is prolonged. The vestibular system responds as before and a compensatory eye movement is produced. If this eye movement were to persist, the eyes would soon reach the limit of gaze, and then all compensation would be lost, but this does not happen. Instead, when the eyes reach a certain deviation of gaze, the pulse generator issues the signal for a saccade in the opposite direction, which puts the eyes on a new target, whereupon the eyes resume their compensatory deviation, and the process is repeated. The result is a back and forth eye movement known as *nystagmus.* It always has components: a *slow phase,* which is the compensatory movement generated by the signal from the labyrinths, and a *fast phase,* which is the saccade in the opposite direction. What triggers the pulse generator to produce the saccades is unknown.

The second thing that happens during prolonged head movements is that the vestibular input declines. As the movement proceeds the cupulae gradually return to their neutral positions and the firing rate in the afferent neurons wanes (p. 25). As a result the eye movements are truly compensatory only during the first second or two of the movement. If no other input were available, eye movements would gradually slow down and eventually stop as the head movement continued.

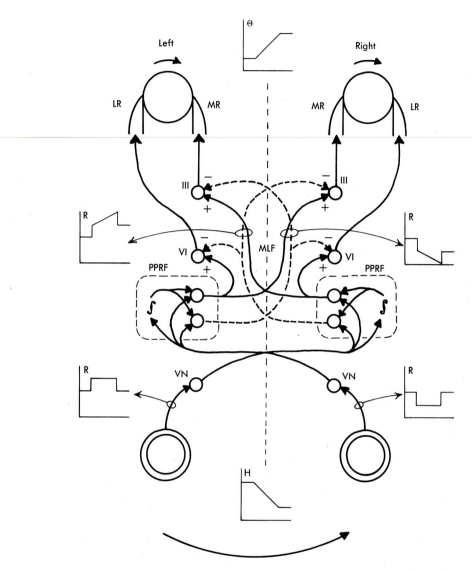

Fig. 2-22. Schematic diagram of the vestibulo-oculomotor pathway and effects of a quick change in head position, *H*, on eye position, *θ*. *LR*, lateral rectus muscle; *MR*, medial rectus muscle; *III*, third (oculomotor) nucleus; *VI*, sixth (abducens) nucleus; *MLF*, medial longitudinal fasciculus; *PPRF*, paramedian pontine reticular formation; *VN*, vestibular nucleus; *R*, neural firing rate.

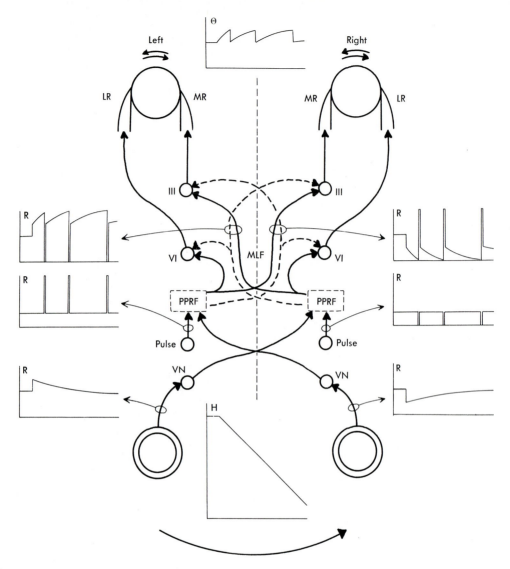

Fig. 2-23. Schematic diagram of the vestibulo-oculomotor pathway and effects of a slow change in head position, *H,* on eye position, *θ.* (Abbreviations are the same as in Fig. 2-22.)

Optokinetic system

Another input is available. Provided that the individual undergoing rotation has his eyes open and is rotating in a stable visual environment, the additional input is provided by the optokinetic system. The way in which the vestibular and optokinetic systems work together is shown in Fig. 2-24. The input to the optokinetic system is not known with certainty; presumably it arises in the retinal movement receptors. It is known, however, that the optokinetic system feeds a signal to

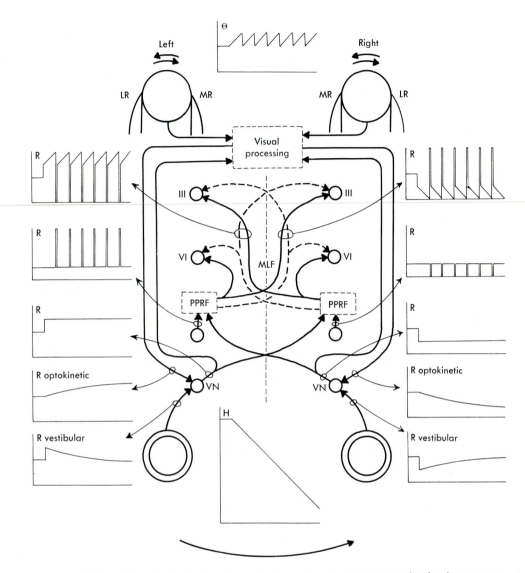

Fig. 2-24. Effects of optokinetic input on the vestibulo-oculomotor response to a slow head movement. (Abbreviations are the same as in Fig. 2-22.)

the vestibular nuclei that is designed to prevent eye movement relative to the environment. This signal is complementary to that provided by the labyrinths.[26-28] The sum of these two signals yields a signal that is exactly proportional to head velocity. The summed signal is fed to the premotor system and then to the motoneurons to produce compensatory eye movement. Note in Fig. 2-24 that the slow phases of the nystagmus are linear and of the same velocity as the head movement. This combined vestibular-optokinetic system is capable of providing compensatory eye movements for a wide range of head movements. When the movements are slow, the input comes mainly from the optokinetic detectors; when they are fast, it comes mainly from the vestibular detectors.

Effects of lesions in the vestibular system. A lesion in the peripheral vestibular system (labyrinth or eighth nerve) causes the tonic input from the damaged side to decrease or, if the lesion is total, to cease altogether. As a result, an asymmetry appears between the inputs from the two sides, which mimics the signal that appears during head rotation. The patient experiences the illusion of motion, or *vertigo,* and he has nystagmus. His nystagmus is primarily horizontal, although it may be horizontal-rotary. It is identical in form to that produced by rotation, with distinct fast and slow phases. The fast phases are toward the intact ear. If the patient's eyes are open and fixating on a stable visual environment, this nystagmus is suppressed by the optokinetic system.

The lesion may be permanent, but the nystagmus (and the vertigo) gradually abate during the ensuing days and weeks. This recovery occurs because of *compensation mechanism* exists within the central nervous system that rebalances persistent vestibular asymmetries.[29,30] The need for such a compensation mechanism is apparent when one considers that the vestibular system operates open-loop, that is, without any feedback. Such a system is vulnerable to random fluctuations in neuron firing rate that must be expected to occur over the person's lifetime. To prevent such fluctuations from causing vertigo and nystagmus, the compensation mechanism continually rebalances persistent asymmetries. The process of compensation is illustrated in Fig. 2-25. Normally tonic neural activity is present in both vestibular nuclei and it is equal on both sides (Fig. 2-25, *A*). When the input from one labyrinth is suddenly interrupted, the spontaneous neural activity in the ipsilateral vestibular nuclei is severely depressed or abolished. As a result, a profound asymmetry occurs between vestibular nuclear activity on the right and that on the left (Fig. 2-25, *B*). The cerebellum responds to this asymmetry by imposing a suppression of activity on the intact side, thus partially rebalancing the asymmetry at a lower level (Fig. 2-25, *C*). Subsequently, spontaneous activity gradually resumes, first on the intact side and then on the deafferented side (Fig. 2-25, *D*). The compensation process is finally complete when the activity of both vestibular nuclear groups is again normal and equal (Fig. 2-25, *E*). This sequence of events occurs over the course of several days or weeks.

Effect of lesions in the optokinetic system. A person with a lesion in his optokinetic system would be expected to have weak or absent nystagmus when viewing a moving visual environment. Depending on the location and extent of the lesion, the defect may be present when the environment moves in all directions or it may be present only when the environment moves in a single direction. There are scant data in the literature concerning the specific effects of an optokinetic lesion because of confusion about what constitutes a stimulus to the optokinetic system. A proper optokinetic stimulus should be a moving visual scene that subtends the patient's entire visual field, and it should give the patient the distinct impression that

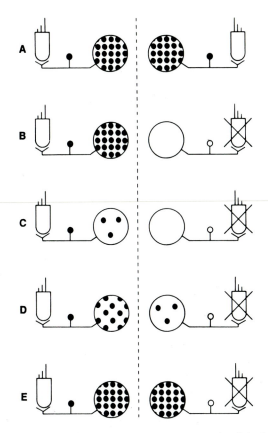

Fig. 2-25. Effect of unilateral labyrinthectomy on spontaneous neural activity in the medial vestibular nuclei. **A,** Normal neural activity. **B,** Immediately after labyrinthectomy. **C,** Two days after labyrinthectomy. **D,** One week after labyrinthectomy. **E,** One month after labyrinthectomy. Density of dots represents level of neural activity. *X* denotes lesion. (A to E, Data from McCabe, B. F., and others: Laryngoscope **82:**381, 1972.)

he, not the visual scene, is moving. The fundamental assumption of the optokinetic system is that *the world never moves,* so it follows that any movement of the entire visual scene is caused by head movement.[26]

Despite this requirement, the "optokinetic" stimulus in current clinical use usually consists of moving black and white stripes that subtend, at most, 30° to 40° of visual angle. Such a stimulus does not provoke a sensation of self-motion. It is in fact a stimulus for the pursuit system (described below), not the optokinetic system.

There is, however, one commonly performed test that clearly evaluates the optokinetic system—the test of visual suppression of inappropriate nystagmus, such as that provoked by a vestibular lesion or by caloric stimulation of the labyrinths. This nystagmus moves the eyes relative to the environment, which normally provokes opposing signals from the optokinetic system. If the optokinetic system is defective, this nystagmus is not suppressed by the optokinetic system. It is as strong with eyes open as it is with eyes closed, and the patient has a defect known as *failure of fixation suppression* of nystagmus. (See Figs. 8-6 and 8-7.)

PURSUIT CONTROL SYSTEM
Function

The task of the pursuit system is to track moving visual targets. It is not entirely clear just how the pursuit system accomplishes this task, but apparently it monitors the rate of target slippage on the retina and sends to the premotor system a signal designed to reduce this slippage to zero.[31]

The pursuit system operates in the same way that the optokinetic system does. There is, however, a basic functional difference between the two. Pursuit eye movements compensate for target motion and optokinetic movements compensate for head motion. In general, movement of small targets against a stable background stimulates the pursuit system and is interpreted as target motion, whereas movement of the entire visual scene stimulates the optokinetic system and is interpreted as self-motion. Sometimes targets are misclassified and a compelling illusion results. For example, if a person is waiting in his car at a stoplight beside a large truck that is inching forward, he may have the sensation that the truck is stationary and he is rolling backward. The illusion is enhanced if he is facing uphill, and it may produce panic-stricken stabs at the brake pedal if another car is parked directly behind. On the other hand, the size of the moving target is not the only factor that determines which system will be stimulated. Other factors, such as the mental set of the patient, also contribute. For example, a patient who is placed inside a moving optokinetic drum often does not experience self-motion, even though the stimulus covers the entire visual field, if he clearly understands that the drum is moving and his chair is fixed.

Pursuit movements, like optokinetic movements, depend on the presence of a moving stimulus. In Fig. 2-26, the upper tracing shows the pursuit movements of a subject following a swinging pendulum bob. The lower tracing shows the same individual trying to make the same movements with his eyes closed. He is unable to make smooth movements; instead he approximates target motion with a series of saccades. Saccades are the only eye movements that can be made at will by most people, although there are a few exceptional persons who can make smooth movements in the absence of a moving stimulus.[32]

Pursuit commands are probably generated in the parieto-occipital visual association areas and reach the ipsilateral PPRF via the internal sagittal stratum. The command signal is proportional to target velocity. It is processed by the premotor system to produce the ramp-step needed to keep the eyes on target.[31] The pursuit system must also send some sort of inhibitory signal to prevent pursuit eye movements from stimulating the optokinetic system when moving targets are tracked against a stable background. The nature of this signal and the neurologic pathways involved are unknown.

Effect of lesions in the pursuit control system. Lesions in the pursuit system produce a distinctive defect known as saccadic pursuit or "cogwheeling." As a

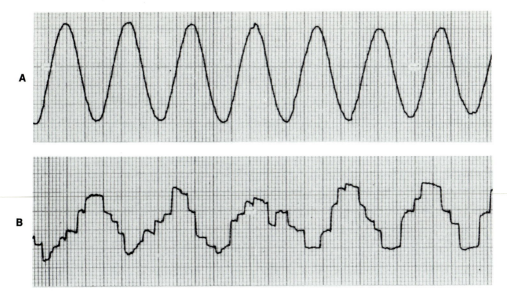

Fig. 2-26. A, Pursuit movement performed while patient is following a moving target. **B,** Pursuit movement performed when patient's eyes are closed while attempting to "follow" an imaginary moving target.

result of the lesion, the gain of the pursuit system is reduced and the eyes cannot keep up with the target, especially when it is moving quickly.[33] The eyes repeatedly slip off the target and then are repositioned on the target by saccades, producing a "notchy" waveform on the ENG tracing. An example is shown in Fig. 6-13.

SACCADIC CONTROL SYSTEM
Function

When an interesting visual target appears in the periphery of the visual field, the eyes can move rapidly to place its image on the foveae. This eye movement is a saccade. It is the quickest eye movement of which the oculomotor system is capable. The speed of a saccade depends on its amplitude; it virtually always exceeds 145° of eye rotation per sec for a small movement and sometimes achieves 700°/sec.[34]

Almost all voluntary shifts of gaze are performed by saccades. One example is the series of eye movements made when reading lines of print (Fig. 2-27). The visual stimulus for saccades seems to be retinal position error, that is, the difference between the actual position of the target on the retina and the desired position (the fovea). Visually evoked saccades are remarkably stereotyped. Once a position error is perceived, there is a 200-msec delay; then the eyes jump to the new position. The agonist eye muscles contract maximally, and the antagonist muscles relax completely in order to make the movement as rapidly as possible. Visual perception is momentarily suppressed during saccades to prevent blurring.[35]

Saccades can also occur in the absence of a visual stimulus. They can be made

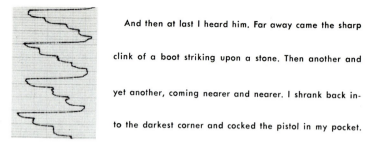

And then at last I heard him. Far away came the sharp

clink of a boot striking upon a stone. Then another and

yet another, coming nearer and nearer. I shrank back in-

to the darkest corner and cocked the pistol in my pocket.

Fig. 2-27. Eye movements performed while patient is reading lines of print. Small rightward saccades fixate on successive groups of words, and large leftward saccades return the eyes to the beginning of the lines.

voluntarily with the eyes closed or in darkness. In addition, saccades periodically interrupt the slow compensatory eye movements generated by optokinetic and vestibular stimuli and thus form the quick phases of the nystagmus associated with these stimuli. The quick phases of nystagmus are indistinguishable from voluntary or visually evoked saccades.[36]

Voluntary saccades are probably generated in the frontal eye fields and descend in the internal capsule to the contralateral PPRF.[37] The output of the saccadic control system is a "desired eye position" signal which is sent to comparator neurons in the burst generator. The saccade is performed only when the "trigger" signal turns off the pause cells. The neural mechanisms underlying the trigger signal are unknown.

Effects of lesions in the saccadic control system. Patients with acute lesions of the frontal cortex have a defect known as *saccadic palsy*.[33] They are unable to make a saccade to the side opposite the lesion. Pursuit movements are intact. When the patient is exposed to optokinetic or vestibular stimulation that calls for nystagmus with fast phases toward the side opposite the lesion, the fast phases are absent and the eyes deviate tonically in the direction of the slow phase. If the opposite hemisphere is intact, saccadic palsy recovers in a matter of days or weeks. A bilateral frontal lesion causes a bilateral saccadic palsy that is permanent.

Another type of saccade defect is *dysmetria*. The "desired eye position" signal is incorrect, so the saccade does not place the eyes on the target. The saccade may be too small or too large. In either case, additional saccades are subsequently generated that bring the eyes successively closer until finally they acquire the target. It appears that a patient with saccadic dysmetria finds the target by trial and error. Examples of saccadic dysmetria are shown in Figs. 6-2 and 6-3.

Another saccadic defect is known as *square wave jerks*. During steady fixation, a spurious saccade throws the eyes off target, a second saccade returns them to the target again, another spurious saccade throws them off, and so on. The ENG waveform of such a defect looks like a square wave. An example is shown in Fig. 5-23.

REFERENCES

1. Burde, R. M.: The extraocular muscles. Part I. Anatomy, physiology, and pharmacology. In Moses, R. A., editor: Adler's physiology of the eye, ed. 6, St. Louis, 1975, The C. V. Mosby Co., pp. 86-122.
2. Burde, R. M.: The extraocular muscles. Part II. Control of eye movements. In Moses, R. A., editor: Adler's physiology of the eye, ed. 6, St. Louis, 1975, The C. V. Mosby Co., pp. 123-165.
3. Fuchs, A. F., and Luschei, E. S.: Firing patterns of abducens neurons of alert monkeys in relationship to horizontal eye movement, J. Neurophysiol. 33:382, 1970.
4. Robinson, D. A.: Oculomotor unit behavior in the monkey, J. Neurophysiol. 33:393, 1970.
5. Bach-y-Rita, P.: Neurophysiology of eye movements. In Bach-y-Rita, P., Collins, C., and Hyde, J. E., editors: The control of eye movements, New York, 1971, Academic Press, Inc., pp. 7-45.
6. Glaser, J. S.: Infranuclear disorders of eye movement. In Glaser, J. S., editor: Neuro-ophthalmology, New York, 1978, Harper & Row, Publishers, pp. 245-284.
7. Robinson, D. A.: Oculomotor control signals. In Lennerstrand, G., and Bach-y-Rita, P., editors: Basic mechanisms of ocular motility and their clinical implications, New York, 1974, Pergamon Press, Inc., pp. 337-374.
8. Pola, J., and Robinson, D. A.: An explanation of eye movements seen in internuclear ophthalmoplegia, Arch. Neurol. 33:447, 1976.
9. Heun, V., and Cohen, B.: Coding of information about rapid eye movements in the pontine reticular formation of alert monkeys, Brain Res. 108:307, 1976.
10. Daroff, R. B., Troost, B. T., and Dell'Osso, L. F.: Nystagmus and related ocular oscillations. In Glaser, J. S., editor: Neuro-ophthalmology, New York, 1978, Harper & Row, Publishers, pp. 220-240.
11. Carpenter, R. H. S.: Cerebellectomy and the transfer function of the vestibulo-ocular reflex in the decerebrate cat, Proc. R. Soc. Lond. Biol. 181:353, 1972.
12. Robinson, D. A.: The effect of cerebellectomy on the cat's vestibuloocular integrator, Brain Res. 71:195, 1974.
13. Zee, D. S., and Robinson, D. A.: A hypothetical explanation of saccadic oscillations, Ann. Neurol. 5:405, 1979.
14. Keller, E. L.: Participation of the medial pontine reticular formation in eye movement generation in monkey, J. Neurophysiol. 37: 316, 1974.
15. Keller, E.: Control of saccadic eye movements by midline brain stem neurons. In Baker, R., and Berthoz, A., editors: Control of gaze by brain stem neurons, New York, 1977, Elsevier North Holland Publishing Co., pp. 327-336.
16. Shults, W. T., and others: Normal saccadic structure of voluntary nystagmus, Arch. Ophthalmol. 95:1399-1404, 1974.
17. Watanuki, K., and Schuknecht, H. F.: A morphological study of human vestibular sensory epithelia, Arch. Otolaryngol. 102: 583-588, 1976.
18. Lindeman, H. H.: Studies on the morphology of the sensory regions of the vestibular apparatus, Adv. Anat. Embryol. Cell Biol. 42:1, 1969.
19. Goldberg, J. M., and Fernandez, C.: Physiology of peripheral neurons innervating semicircular canals of the squirrel monkey: I. Resting discharge and response to constant angular accelerations, J. Neurophysiol. 34:635, 1971.
20. Flock, A.: Sensory transduction in hair cells. In Lowenstein, W. R., editor: Handbook of sensory physiology, Vol. 1, New York, 1971, Springer Publishing Co., Inc., pp. 239-352.
21. Melville-Jones, G.: Transfer function of labyrinthine volleys through the vestibular nuclei. In Brodal, A., and Pompeiano, O., editors: Basic aspects of central vestibular mechanisms: progress in brain research, Vol. 37, New York, 1972, Elsevier North Holland Publishing Co., pp. 139-156.
22. Young, J. H.: Analysis of vestibular system responses to thermal gradients induced in the temporal bone, thesis, University of Michigan, Ann Arbor, 1972.
23. Brodal, A.: Anatomy of the vestibular nuclei and their connections. In Kornhuber, H. H., editor: Handbook of sensory physiology, Vol. 6, New York, 1974, Springer Publishing Co., Inc., pp. 239-352.
24. Ito, M.: Neural design of the cerebellar motor control system, Brain Res. 40:81, 1972.
25. Precht, W.: Vestibular mechanisms, Ann. Rev. Neurosci. 2:265, 1979.
26. Robinson, D. A.: Vestibular and optokinetic symbiosis: an example of explaining by modelling. In Baker, R., and Berthoz, A., edi-

tors: Control of gaze by brain stem neurons, New York, 1977, Elsevier North Holland Publishing Co., pp. 49-58.

27. Waespe, W., and Henn, V.: Neuronal activity in the vestibular nuclei of the alert monkey during vestibular and optokinetic stimulation, Exp. Brain Res. **27:**523, 1977.

28. Waespe, W., and Henn, V.: Vestibular nuclei activity during optokinetic after nystagmus (OKAN) in the alert monkey, Exp. Brain Res. **30:**323-330, 1977.

29. Precht, W.: Characteristics of vestibular neurons after acute and chronic labyrinthine destruction. In Kornhuber, H. H., editor: Handbook of sensory physiology, Vol. 6, New York, 1974, Springer Publishing Co., Inc., pp. 451-462.

30. McCabe, B. F., Ryu, J. H., and Sekitani, T.: Further experiments on vestibular compensation, Laryngoscope **82:**381, 1972.

31. Robinson, D. A.: The mechanics of human smooth pursuit eye movement, J. Physiol. **180:**569, 1965.

32. Deckert, G. H.: Pursuit movements in the absence of a moving visual stimulus, Science **143:**1192, 1964.

33. Daroff, R. B., and Troost, B. T.: Supranuclear disorders of eye movements. In Glaser, J. S. editor: Neuro-ophthalmology, New York, 1978, Harper & Row, Publishers, pp. 202-218.

34. Boghen, D., and others: Velocity characteristics of normal human saccades, Invest. Ophthalmol. **13:**619, 1974.

35. Zuber, B. L., and Stark, L.: Saccadic suppression: elevation of visual threshold associated with saccadic eye movements, Exp. Neurol. **16:**65, 1966.

36. Ron, S., Robinson, D. A., and Skavenski, A. A.: Saccades and the quick phases of nystagmus, Vision Res. **12:**2015, 1972.

37. Brocher, J. M.: The frontal eye field of the monkey, Int. J. Neurol. **5:**262, 1966.

Chapter 3

The ENG laboratory

Once a physician decides to include ENG in his practice, he is faced with the task of setting up a laboratory. Ordinarily, the first step is to decide which of the ENG tests he wishes to perform, because this decision will determine how his laboratory is to be arranged. We will assume here that the physician wishes to conduct seven tests—the gaze test, the saccade test, the tracking test, the optokinetic test, the positional test, the Hallpike maneuver, and the caloric test—which comprise the basic ENG examination.

THE ROOM

After the tests have been selected, the next task is to find a suitable room to house the laboratory. A layout suitable for performing the basic ENG examination is shown in Fig. 3-1. The dimensions of this room are 10 ft × 14 ft, which is probably the smallest size that could be used. The patient is placed in the middle of the room with the visual display in front of him and the major ENG equipment (nystagmograph and caloric irrigator) behind. A sink is in one corner, and a cabinet for storage of supplies is in the other.

The room that will be the ENG laboratory should possess several special features. First, it should be located away from devices such as x-ray and diathermy machines, which emit large amounts of electrostatic and electromagnetic energy that might interfere with ENG recording. Second, air-conditioning and ventilation should be adequate to keep the room cool and dry. Consideration must be given to the fact that a caloric water bath generates considerable heat and moisture, the archenemies of electronic apparatus. Furthermore, if the room becomes too warm, the patient will begin to perspire, and satisfactory eye movement recordings cannot be obtained from him. Third, it may be desirable to lightproof the laboratory. Although most basic ENG procedures are conducted in dim light, there may be instances in which it is better to record eye movements with the patient's eyes open in absolute darkness. If such a condition is desired, the lightproofing of the laboratory must be perfect, because even a pinpoint of light is sufficient to provide visual fixation.

ELECTRICAL SAFETY

The patient who undergoes ENG testing is surrounded by devices that are powered by electric current obtained from the wall outlets, and he is placed in di-

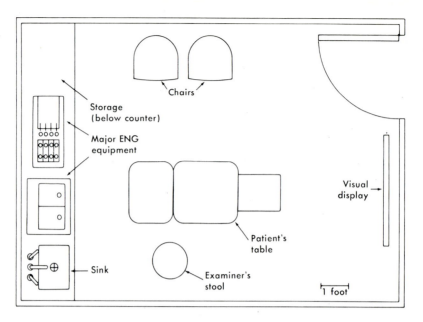

Fig. 3-1. Sample layout of an ENG laboratory.

rect contact with one of those devices, the nystagmograph, by low-impedance connections, the electrodes. Therefore, special precaution must be taken to ensure that he does not receive an electric shock.

Each electrical outlet in the wall of the laboratory contains three metallic conductors. Two of them carry the electric current: one is the *"neutral" wire,* which is attached directly to ground (usually through a water pipe buried deeply in the earth); the other is the *"hot" wire,* which is at a voltage of 110 v relative to the grounded wire (Fig. 3-2). Current flows through the internal circuitry of the device that is plugged into the outlet when the operator closes the switch that completes the circuit between the hot wire and the neutral wire. The third wire is called the *"ground" wire;* it is attached to ground and to the exposed conductive surfaces of the device plugged into the outlet. (This system is standard in the United States and Canada, but a different system may be in use elsewhere.)

The purpose of the ground wire is to offer a direct pathway to ground for any stray electric currents that leak from the internal circuitry of the device onto its exposed surfaces. The way in which the system is supposed to work is illustrated in Fig. 3-2, *A;* the insulation has worn away from the power cord of a motorized examining table, allowing the hot wire to touch the metal frame. Even though the switch is open, a circuit to ground is completed via the ground wire.

According to Ohm's law, the amount of electrical current (I) that will flow through any conductive pathway is directly proportional to the applied voltage (E)

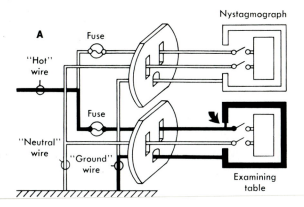

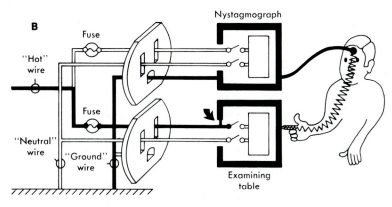

Fig. 3-2. Consequences of an accidental short circuit (arrow) between the "hot" wire and the metal frame of the patient's examining table. **A,** When the frame is grounded, the ground wire completes the circuit and blows the fuse. **B,** When the frame is not grounded, it remains at 110 v and poses a shock hazard to the grounded patient who touches it. The solid black lines indicate the involved current pathways.

and inversely proportional to the pathway's impedance (Z), or E/Z = I. The electrical outlet is a constant voltage source; that is, it delivers 110 v regardless of the current drawn. Thus the amount of current that will flow through the ground wire is determined solely by the impedance it offers between the hot wire and ground. If properly installed and maintained, the ground wire offers very little impedance, typically less than 1 ohm. When this value is substituted into Ohm's equation, then 110 v/1 ohm = 110 amps. Thus a massive current surges through the ground wire, which instantly blows the fuse and disconnects the entire circuit from electric power.

Fig. 3-2, *B*, shows what can happen if the metal frame is not grounded. In this case, there is no pathway to ground; therefore, the circuit is not completed and the frame remains at the full-line potential of 110 v. If the patient then touches the frame, his body completes the circuit, because he is attached directly to ground via the ENG electrode and the nystagmograph. The amount of current that will flow through his body depends upon its impedance, most of which is contributed by the skin. Skin impedance is quite variable, but dry skin usually has an impedance of 100,000 ohms. When this value is substituted into Ohm's equation, then 110 v/100,000 ohms = 0.001 amp, or 1 ma. Thus the patient with dry skin will experience only a slight tingle, because 1 ma is barely above the threshold of perception.[1] He will receive a much stronger current, however, if his hands are moist. The impedance of moist skin can be as low as 1,000 ohms. (The technician has already established a low-impedance connection between the patient and ground via the ENG electrode.) Substituting this value into Ohm's equation, one sees that the patient with moist hands will receive 110 v/1,000 ohms = 0.110 amps, or 110 ma. This amount of current is sufficient, under certain circumstances, to produce ventricular fibrillation.[1]

Note that the current that flows through the patient would not be enough to blow the fuse. Fuses are placed in the circuit to disconnect power only when current becomes strong enough to heat up the electrical wires and create a fire hazard. Fuses in electrical devices are usually designed to blow when current exceeds 5 amps; those in power distribution boxes will blow when current exceeds 20 amps. Such current levels far exceed those sufficient to produce a lethal electric shock. The frame of the examining table in Fig. 3-2, *B*, would have remained "hot" until someone unplugged it.

Clearly the matter of electrical safety merits serious attention. Each wall outlet should be checked to ensure that its polarity is correct and that its ground connection is intact. Electric devices placed in the ENG laboratory should be checked to ensure that the pathway from chassis to ground prong is intact, and devices without such a prong should be banned. Current leakage from internal circuits to exposed conductive surfaces should be checked. These checks should be performed before the devices are used and periodically (usually at 90-day intervals) thereafter. Few users of ENG have the expertise or the specialized equipment needed to perform proper electrical safety checks of their laboratories by themselves. Ordinarily, they must entrust this work to specialists in their hospital's biomedical electronics department or to other qualified personnel.

Users of ENG are obligated, of course, to follow the standards for electrical safety adopted by their own hospital. Most hospitals in the United States now follow the standard developed by the Association for the Advancement of Medical Instrumentation (AAMI).[2] It is commonly referred to as "the AAMI leakage current standard" and is generally consistent with standards set or proposed by other standards — writing bodies, such as the Underwriters' Laboratories (UL) and the National Fire Protection Association (NFPA). This standard has been adopted by the American National Standards Institute (ANSI), where it is designated

ANSI/AAMI SCL 1/78. Electrical safety standards adopted by hospitals in other countries are not necessarily the same as those followed in the United States. ENG users in other countries should consult with a biomedical engineer regarding the standards in use in their own country.

Users of ENG should also be aware that, even though their laboratory has been proved safe for patients wearing electrodes, it still may be hazardous for the patient who wears a low-impedance connection to his heart, such as an intracardiac catheter or temporary pacemaker with exposed leads. Minute electric currents, perhaps as low as 20 μamps, across those leads might be hazardous to this patient, because these currents would be conducted directly to cardiac tissue. Such a patient should not be allowed to enter the ENG laboratory unless special precautions have been taken to protect him. Patients with implanted permanent pacemakers are at no risk.

EYE MOVEMENT RECORDING SYSTEM

A steady potential, known as the corneoretinal potential, exists between the front and the back of the eye. The size of this potential is normally at least 1 mv, and its axis coincides roughly with the visual axis (Fig. 3-3). The corneoretinal potential produces in the front of the head an electrical field, which changes in its orientation as the eyes move. When a pair of electrodes is placed on the skin, as shown in Fig. 3-3, a voltage that is nearly a linear function of eye position appears between them. The changes in voltage produced by eye displacement are small, only about 20 μv/degree.

The task performed by the eye movement recording system is a demanding one. The system must detect the minute voltage changes produced by eye displacement, amplify them by about 20,000 times without distortion, and display them in a usable form. At the same time, the recording system should not accept extraneous voltages from other sources or add any of its own.

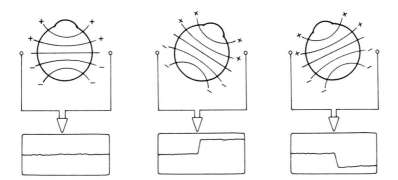

Fig. 3-3. Effect of changes in the orientation of the corneoretinal potential on the voltage between recording electrodes.

Electrodes

The purpose of the recording electrodes is to detect the voltage changes generated by eye movements and present them to the input terminals of the nystagmograph. The type of electrode used almost universally for this purpose consists of an *Ag/AgCl pellet* mounted in a plastic cup that holds it away from the skin (Fig 3-4). The space between the pellet and the skin is filled with electrolytic paste, and the electrode is attached tm the skin by a doughnut-shaped ring cut from adhesive-coated plastic tape. A flexible lead wire connects the pellet to the input terminals of the nystagmograph.

This type of electrode is widely used because it does not polarize as readily as other metal electrodes do.[3] All metal electrodes polarize to some extent, because a metal tends to discharge ions into solution when the metal is placed in contact with an electrolyte, and the ions in the solution tend to combine with the metal. This reaction is complex, but in its simplest form, it can be represented as a layer of ions tightly bound to the surface of the metal and an adjacent layer of oppositely charged ions in the solution (Fig. 3-4, *inset*).

This double layer of ions is troublesome for several reasons. First, it causes the electrode-electrolyte interface to acquire a potential, known as the *half-cell potential*. This potential would be of little concern if it were stable, but unfortunately it is unstable. It continuously fluctuates in a random manner, producing an artifact in the ENG tracing that is indistinguishable from the signal produced by eye movements. To make matters worse, this electrical double layer is sensitive to movement. Any mechanical disturbance of the electrode-electrolyte interface upsets the double layer, causing changes in voltage that are often very large in comparison with those produced by eye movements.

The double layer also endows the electrode-electrolyte interface with an impedance. This impedance is undesirable because it attenuates and distorts the eye movement signal unless a nystagmograph with a high input impedance is used. (Input impedance of nystagmographs is discussed on pp. 51-52.

The Ag/AgCl electrode displays the highest stability and lowest impedance of

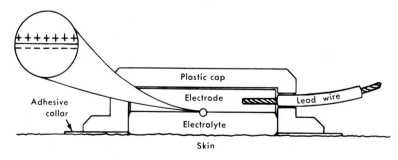

Fig. 3-4. A typical recording electrode. *Inset,* electrical double layer at the electrode-electrolyte interface.

any readily available electrode and is therefore desirable for eye movement recording. Moreover, the design of the particular Ag/AgCl electrode shown in Fig. 3-4 is relatively insensitive to movement because the electrode-electrolyte interface is located a short distance away from the skin and is somewhat protected from mechanical displacement.

The electrolyte used under the electrode is designed to overcome the electrical impedance of the dry epidermis. As it soaks through the epidermal layers, it establishes a conductive pathway between the body fluids and the surface of the electrode. Any of the electrolyte compounds commonly used for recording electrocardiograms are suitable for eye movement recordings.

Nystagmographs

A nystagmograph amplifies and displays the minute voltage chances detected by the recording electrodes. Most nystagmographs are composed of the elements shown in Fig. 3-5. The first element is the signal conditioner, which extracts the useful parts of the incoming signal. The signal conditioner performs this function primarily by filtering out parts of the signal that the operator judges to be extraneous. The preamplifier boosts the voltage of the input signal, and the power amplifier converts this voltage into the electric current needed to drive the output transducer. The output transducer is a galvanometer that transforms electric current into pen position. As the recording paper is moved under the pen by the paper drive, a permanent record of the input signal is traced upon it. The power supply takes electric current from the wall outlet and delivers it in appropriate form to the other components.

Various kinds of biomedical recorders are suitable for recording eye movements. Devices called "nystagmographs" by their manufacturers are designed specifically for this task, but general purpose recorders can be used, provided they meet the specifications listed in Table 3-1 (p. 59). Devices specifically designed to record other biologic signals, such as the electroencephalogram or electrocardiogram, usually must be modified before they will yield acceptable eye movement tracings.

The specifications of a nystagmograph are quantitative indices of its performance based on certain important factors. They are found, in fine print, in the man-

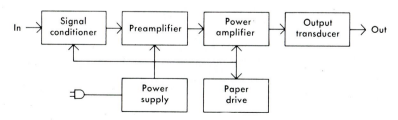

Fig. 3-5. Elements of a typical nystagmograph.

ufacturer's brochure and help the prospective buyer make comparisons among various models. The most important factors for which specifications are generally available are the following.

Type of input. Two types of input are used in biologic amplifiers: *single-ended* and *differential* (Fig. 3-6). In an amplifier with a single-ended input, one of the electrode leads is attached directly to ground; therefore, any voltage detected by the other (active) electrode with respect to ground potential is amplified. In a differential amplifier, both input leads are active, and only the potential difference between them is amplified. Only differential amplifiers are used for eye movement recording. Single-ended amplifiers are unsuitable because the single active electrode picks up any voltage not at ground potential, including ground-referred interference from power lines. On the other hand, the differential amplifier is relatively insensitive to electrical interference signals because such signals are present simultaneously in both input leads and, therefore, do not produce a potential difference between them.

Common mode rejection ratio. The ease with which a differential amplifier rejects electrical signals that are present simultaneously in the two input leads directly affects its ability to produce noise-free tracings. Some amplifiers perform better than others in this respect, and the common mode rejection ratio is an index of that performance. The manufacturer determines the common mode rejection ratio as follows[4]: First, he applies a differential voltage of known frequency and intensity to the input terminals of the amplifier (Fig. 3-7, *A*) and measures the output voltage. Then he applies a common mode voltage of the same frequency (Fig. 3-7, *B*) and increases its intensity until it yields the same output voltage as before. (Both input voltages must be within the operating range of the amplifier.) The ratio of these two input voltages is the common mode rejection ratio. For example, if a 1-v common mode input signal is required to produce the same output as a 10 μv differential input signal, the common mode rejection ratio is 100,000:1, or 100 dB. The higher this ratio, the better the ability of the amplifier to reject electrical interference. In practice, a common mode rejection ratio of 80 dB or higher for 60 Hz is adequate.

The user should be aware that the manufacturer reports the common mode rejection ratio obtained under the most favorable conditions, that is, when the impedances of the two active input leads are equal. When the impedances of the two leads are unequal, common mode voltages are stronger at one input terminal than at the other, producing a differential voltage that is accepted by the amplifier. In practice, unequal impedances in the two input leads are usually caused by the fact that one electrode-skin junction of a pair has a low impedance and the other has a high impedance because it was applied improperly. When this happens, the common mode rejection ratio of the amplifier falls below its rated value, and electrical interference appears in the record.

Input sensitivity. The input sensitivity of the nystagmograph is the input voltage necessary to produce a given pen deflection, usually 1 mm, when the gain is

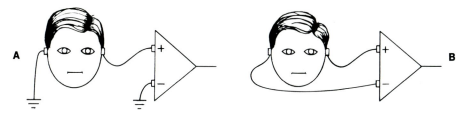

Fig. 3-6. Types of input in biologic amplifiers. A, Single ended. B, Differential.

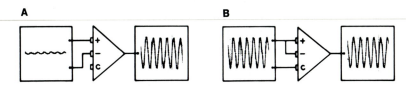

Fig. 3-7. Method of determining common mode rejection ratio. A, Applying a differential voltage. B, Applying a common mode voltage.

adjusted to maximum. When performing ENG procedures, the operator ordinarily adjusts the gain so that 1° of eye movement yields 1 mm of pen deflection. Because 1° of eye movement produces a voltage change of approximately 20 μv for most individuals, the gain ordinarily will be set to yield 20 μv/mm. As a rule of thumb, it is desirable for the maximum sensitivity of the nystagmograph to be approximately ten times the sensitivity that is routinely used; thus maximum input sensitivity should be approximately 2 μv/mm. If it is much lower, the nystagmograph may not be sensitive enough; if it is higher, the extra sensitivity will rarely be utilized.

The nystagmograph should have an attenuator control, which allows the operator to adjust the input sensitivity in discrete steps over the entire operating range of the instrument. In addition, a continuously variable (vernier) attenuator should be available for making delicate sensitivity adjustments over a range of approximately 10 dB.

Noise level and drift. Instability of the electronic components of the nystagmograph can cause high-frequency noise and low-frequency drift to appear as artifacts in the tracing; however, in modern nystagmographs, this source of artifact is almost negligible. When the input leads are shorted together, noise level should be less than 10 μv (peak to peak) and drift should be less than 50 μv (peak to peak) per hour.

Input impedance. The nystagmograph measures the voltage that appears between the two electrode leads, and like any voltage-measuring instrument, it should possess a high input impedance. If the nystagmograph has an input impedance that is too low, excessive current flows into, or "loads," the input circuit and thereby produces attenuation and distortion of the voltage signals generated by eye movement.[3]

A schematic diagram of the electrical circuit formed by the tissue and electrodes is shown in Fig. 3-8, *A*. This circuit consists of a voltage generator (the corneoretinal potential), tissue impedances, and electrode impedances. The sum total of these impedances (Z_0) is typically 10,000 ohms or less. If a voltage-measuring device having an infinite input impedance were connected to the tips of the electrode leads, it would yield a voltage reading of E_0. However, when a nystagmograph is connected to the leads (Fig. 3-8, *B*), it yields a smaller voltage reading, E_i, because it has a finite input impedance, Z_i, which "loads" the circuit and attenuates the voltage signal. The amount of voltage attenuation that is caused by nystagmograph input impedance can be determined from the formula,

$$\frac{E_i}{E_0} = 1 - \frac{Z_0}{Z_0 + Z_i}$$

By inserting various values of Z_i into the equation, one can see the effect of input impedance on the voltage signal produced by eye movements. For example, if both Z_0 and $Z_i = 10,000$ ohms, then $E_i/E_0 = 0.50$, which means that the nystagmograph would detect only half of the available voltage. To compensate, one would have to double the gain, which of course would double the level of noise in the tracing.

To avoid significant loss of voltage, Z_0 should be as small as possible, and Z_i should be as large as possible. As a rule of thumb, Z_i should be approximately 100 times larger than Z_0. If $Z_0 = 10,000$ ohms, and $Z_i = 1$ megohm, then $E_i/E_0 = 0.99$. In other words, a nystagmograph having an input impedance of 1 megohm would cause a voltage attenuation of only 1 percent, which is insignificant.

The other consequence of low input impedance, signal distortion, is difficult to describe simply. It depends on signal frequency and amplitude, among other factors, and consists primarily of electronic differentiation, with the result that low-frequency components of the signal are lost. Signal distortion is insignificant in eye movement recordings when input impedance of the nystagmograph is greater than 1 megohm.

Frequency filters. Even when a well-designed nystagmograph is used under perfect recording conditions, it is sometimes impossible to eliminate all unwanted electrical signals, especially those at the power line frequency of 60 Hz. Fortunately, most of the voltage produced by nystagmus lies below 30 Hz[5]; therefore, it is possible to alleviate the problem of noise somewhat by incorporating a *low-pass filter* into the nystagmograph. Fig. 3-9 shows a 30-Hz low-pass filter and its effect on sine wave signals of various frequencies. This filter is called a 30-Hz filter because it attenuates voltages at that particular frequency by 3 dB (30 percent). Note that this filter does not pass all signals below 30 Hz or reject all signals above 30 Hz. A perfect 30-Hz filter would do so, but filters on nystagmographs are not perfect. These filters possess a characteristic called *rolloff,* which is an

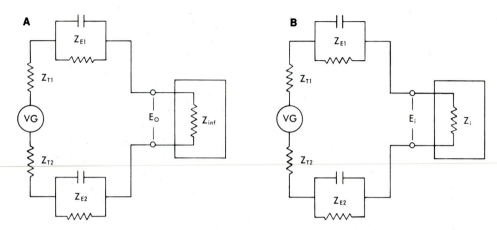

Fig. 3-8. A, Input circuit consisting of a voltage generator *(VG)*, tissue impedances *(Z_{T_1}* and *Z_{T_2})*, and electrode impedances *(Z_{E1}* and *Z_{E2})*. The sum of these impedances is *Z_0*. If a device having an infinite imput impedance *(Z_{inf})* were connected to the electrode leads, it would measure a voltage of *E_0*. **B,** Same input circuit. When a device having a finite input impedance *(Z_i)* is connected to the electrode leads, it measures a lower voltage, *E_i*.

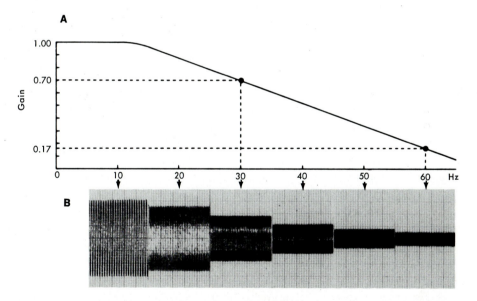

Fig. 3-9. A, Frequency characteristics of a 30-Hz low-pass filter with 12 dB per octave roll-off. **B,** Effect of this filter on sine waves of various frequencies. It attenuates the 30-Hz signal by 3 dB (30 percent). It attenuates the 60-Hz (1 octave higher) signal by an additional 12 dB (75 percent), reducing its strength to 17 percent of the input signal.

expression of the rate at which progressively higher frequencies are attenuated. The rolloff of the filter shown in Fig. 3-9 is approximately 12 dB/octave; in other words, each time the signal frequency is doubled, the filter attenuates it by 12 dB. The greater the rolloff, the better the filter. The choice of a frequency filter always represents a compromise between fidelity and convenience. A 30-Hz low-pass filter is a reasonable compromise for eye movement recording.

The issue of *high-pass filters* in ENG is still one of debate. Most nystagmographs can operate without a high-pass filter and thus are able to measure steady, or DC potentials. A DC nystagmograph is required for faithful recording of eye position. Unfortunately, DC recording is demanding, because factors other than eye position can affect the DC potential between electrodes. One factor is the fluctuation that occurs in the potential at the electrode-electrolyte junction. Other factors are slow changes that take place in the potentials at the electrolyte-skin junction and in the skin itself. These potentials cannot be entirely eliminated, and sometimes they produce continuous and unpredictable changes in the recorded DC potential, requiring frequent repositioning of the pen to keep it from drifting to the edge of the recording channel. To avoid drift, some clinicians employ a high-pass filter, which rejects these steady or extremely slowly changing potentials. By rejecting low-frequency signals, these filters automatically keep the pen positioned near the center of the recording channel and considerably simplify the recording of nystagmus in the presence of slowly changing potentials. As a consequence, however, any information regarding absolute eye position is lost. A nystagmograph operating with a high-pass filter is said to be operating in the AC mode.

High-pass filters, like low-pass filters, are designated in terms of the particular frequency that they attenuate by 3 dB. Fig. 3-10, *A*, shows the characteristics of a 0.053-Hz high-pass filter. High-pass filters can also be designated in terms of their responses to a sudden change in DC potential. They cause the response to such a change to decay at an exponential rate back toward the baseline (Fig. 3-10, *B*). The time, in seconds, that it takes for the output of the amplifier to decay to 37 percent of its original value is known as the *time constant* of the filter. The time constant of the 0.053-Hz filter shown in Fig. 3-10 is 3 sec.

A modern nystagmograph should be capable of recording without a high-pass filter; that is, it should be capable of recording in the DC mode. In addition, it should be capable of operating in the AC mode, that is, it should have one or more high-pass filters that can be selected if DC recording is untenable. A 0.053-Hz (3-sec time constant) high-pass filter is suitable for recording eye movements.

Zero suppression. When recording using the DC mode, one needs a control on the nystagmograph that introduces an offset voltage to compensate for any steady DC voltage that is present between the electrodes. This control enables the user to position the pen in the middle of the recording channel. To be adequate, the offset voltage should be at least ± 10 times full-scale pen deflection.

Number of recording channels. Two recording channels are needed: one for

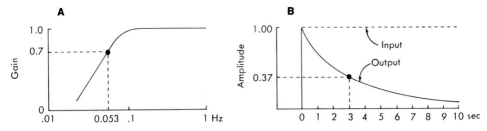

Fig. 3-10. A, Frequency characteristics of a 0.053-Hz (3 sec time constant) high-pass filter. **B,** Effect of this filter on a sudden DC potential change.

horizontal eye movements and one for vertical eye movements. However, a third and fourth channel would be desirable for several reasons. They might be used to make separate recordings from each eye. They might also serve as spares, to be used in case one of the other channels becomes defective. One might also use the additional channels to provide tracings proportional to eye velocity. When the nystagmograph is set up in this fashion, eye position is recorded on one channel. This signal is fed via an internal connection to another channel, which electronically differentiates the eye position signal, yielding a pen deflection proportional to eye velocity. Some clinicians consider the eye velocity recording useful because this measure is commonly used as an index of nystagmus response strength following the caloric irrigation. However, it should be noted that the eye velocity signal is useful only when the eye position signal is extremely noise-free. Electrical interference usually has a prominent high-frequency component, which, when differentiated, yields a high-amplitude signal that sometimes makes the tracing of eye velocity uninterpretable. Examples of differentiated tracings derived from noise-free and from noisy eye position signals are illustrated in Fig. 3-11.

Write-out system. The writing pens on most nystagmographs are heated styli that burn black lines onto the plastic coating of special recording paper. An alternative, offered on only a few nystagmographs, is an ink system in which the ink flows from writing pens either by capillary action or by a pressure system. Both types of write-out systems yield satisfactory tracings. The advantage of the heated stylus system over the ink system is convenience; an ink system can be somewhat messy if not handled properly, and it must be periodically serviced to keep the ink pens flowing, especially when they are not in everyday use. One disadvantage of the heated stylus system is that the paper is more expensive than the paper used in capillary ink systems (although the cost of pressurized ink paper is comparable). Another disadvantage is that tracings made by the heated stylus system are harder to store. Heated stylus paper is available only in difficult-to-handle rolls, whereas ink paper comes in prefolded sheets that can be stored flat in the patient's chart.

When considering write-out systems, one should distinguish between curvilinear and rectilinear recording. The output transducer on all nystagmographs is a

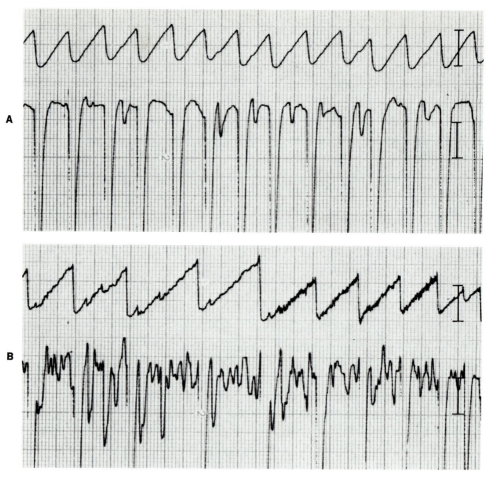

Fig. 3-11 A, Noise-free eye position signal (upper tracing) and differentiated signal (lower tracing). **B,** Noisy eye position signal (upper tracing) and differentiated signal (lower tracing). Paper speed: 10 mm per sec. Calibrations: 10° for eye position signals and 10° per sec for differentiated signals.

galvanometer, which rotates, causing the writing pen, which is attached to the rotating galvanometer coil, to describe an arc. When the pen moves in an arc, the recording is said to be curvilinear. However, a curvilinear tracing is undesirable, because one cannot easily measure important parameters of eye movement, such as eye velocity, and one cannot easily compare simultaneous recordings on more than one channel. These difficulties are avoided when the pen moves back and forth in a direction perpendicular to the movement of the recording paper. This type of pen movement is said to be rectilinear. The rectilinear tracing is compared with the curvilinear tracing in Fig. 3-12.

Because a rectilinear tracing is preferred in ENG recording, most nystagmographs incorporate some sort of mechanism to produce a rectilinear output. When

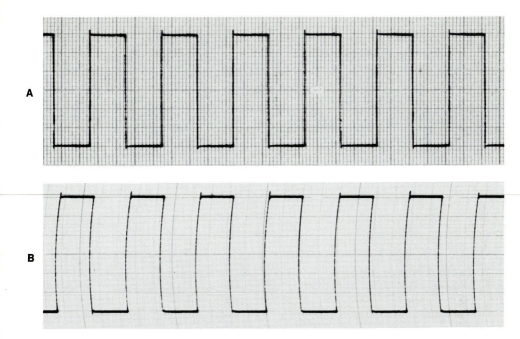

Fig. 3-12. A, Rectilinear tracing. **B,** Curvilinear tracing. (Courtesy David Clark, Ph.D.)

a heated stylus is used, a rectilinear output is achieved in the manner shown in Fig. 3-13. The stylus consists of an inch-long hot wire; the paper moves over the sharp edge of a "forming bar," and the stylus makes contact with the paper only on this edge. As the stylus moves in an arc, a straight line is written. Note that this method of achieving a rectilinear tracing causes a slight underestimate of the amplitude of the recorded signal: true pen displacement is represented by the arc, A, whereas the recorded signal is represented by a straight line, Y. However, this error is not large enough to be significant. Ink pens are converted to rectilinear recording by a special mechanical linkage, which produces the same underestimation of pen displacement as the heated stylus rectilinear system. There is a type of ink recorder that provides a true rectilinear output by moving the recording pen along a slide wire, but the frequency response of this type of recorder is much too low for eye movement recording.

Virtually all nystagmographs have a recording channel width of 40 or 50 mm and a paper speed of 10 mm/sec. In some nystagmographs, other paper speeds may be selected.

Accessories. A timer that produces a signal at 1-sec intervals at the edge of the recording paper is useful, especially if chart speeds other than the standard 10 mm/sec are used.

An arrangement that automatically produces a mark at the edge of the recording channel while the water is flowing during the caloric irrigation is indispensable.

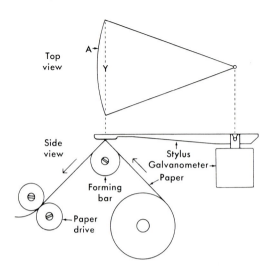

Fig. 3-13. Method of obtaining a rectilinear output with a heated stylus system.

Such an automatic mark is needed for proper evaluation of the caloric response.

Electrical safety. The nystagmograph must meet the criteria for electrical safety that have been adopted by the governing body of one's own hospital. Ordinarily, one would consider this matter in consultation with specialists from the biomedical electronics department of the hospital. One may also wish to read *Safe Current Limits for Electromedical Apparatus*.[2]

Summary. A summary of important specifications for nystagmographs is provided in Table 3-1. This list is by no means exhaustive. A number of other specifications, such as overload protection, amplitude linearity of the output transducer, or stability to line voltage changes, might be considered; however, these specifications are satisfactorily met by virtually all nystagmographs currently on the market.

Data reduction devices

The nystagmograph routinely used to perform the basic ENG procedures produces tracings of eye position and, if electronic differentiators are used, additional tracings of eye velocity. Certain laboratories, especially those that combine routine clinical testing with teaching and research activities, may require automatic data reduction devices to analyze these tracings. A wide variety of such devices are available. There is, for example, at least one device currently available that automatically computes and displays the peak slow-phase eye velocity following the caloric irrigation. Devices such as this must be quite sophisticated if they are to yield accurate data when used on recordings that contain artifacts. Eye velocity is relatively simple to compute when the nystagmus tracings are noise-free but considerably more difficult when they are noisy. In any case, it seems that automatic data reduction devices are necessary only when large amounts of data are

Table 3-1. Specifications for nystagmographs

Parameter	Requirements
Type of input	Differential
Common mode rejection ratio	80 dB or more at 60 Hz
Input sensitivity	2 μv or less per millimeter, with coarse step attenuator and fine vernier attenuator
Noise level	10 μv (peak-to-peak) or less
Drift	50 μv (peak-to-peak) or less per hour
Input impedance	1 megohm or more
Frequency filters	
Low-pass	30 Hz (-3 dB), with rolloff of 12 dB or more per octave above 30 Hz
High-pass	DC and 0.053 Hz (-3 dB), selected by operator
Zero suppression	$\pm$10 times full-scale pen deflection or more
Number of recording channels	2 or more
Write-out system	Heated stylus or ink rectilinear 50-mm channel width 10 mm/sec paper speed
Accessories	Timer signals at 1-sec intervals Contact-closure event marker
Electrical safety	Conforms to standards of one's hospital

generated in the laboratory. The data reduction required when one is performing routine clinical procedures is fairly simple and is not especially time consuming, even when done manually. Other data storage and readout devices, such as tape recorders and oscilloscopes, may be desirable. These devices may be useful when multiple copies of particular tracings are needed for teaching purposes, but again they would probably be infrequently used in routine clinical work.

VESTIBULAR STIMULATORS

In the basic ENG examination, vestibular stimulation is provided by a caloric irrigator. The clinician who intends to purchase a caloric irrigator faces a dilemma: Should he purchase a water caloric or an air caloric irrigator? Water caloric irrigators have been used successfully for many years. Air caloric irrigators are new; they offer certain advantages over water caloric irrigators; however, in the minds of many clinicians, they are as yet unproved.

Water caloric irrigator

The water caloric irrigator is used to deliver the vestibular stimulus in the caloric test. The caloric stimulus is now fairly well standardized; it consists of 250 cc of water delivered to the external auditory canal within 30 sec. Each ear is irrigated twice, once with water at 30° C and once more with water at 44° C. The irrigator itself consists of three components: a pair of water reservoirs to hold the water at the proper temperatures (one at 30° C and the other at 44° C), a calibrated delivery system to deliver the proper amount of water to the ear, and a timing system to control the duration of water flow. In addition, the irrigator should deliver a signal (usually a contact closure) to the event marker of the nystagmograph during the irrigation.

When selecting a caloric irrigator, the clinician should consider the accuracy with which it regulates the temperature of the water that flows from the irrigating tip. In attempting to regulate water temperature, the designer of caloric irrigators faces two problems. His first problem is to maintain constant water temperatures in the reservoirs. This problem has been easily solved with temperature sensors located in the reservoirs themselves; thus most irrigators are able to regulate water temperatures in the reservoirs to within 0.1° C. The second problem is maintaining the water at the correct temperature as it flows through the delivery system, and this problem has been more difficult to solve. Both irrigating temperatures are above the ambient temperature of 22° C; thus, the water loses heat in its journey from the reservoir to the irrigating tip, unless steps are taken to prevent this. Fig. 3-14, *A*, shows the temperature of the water that leaves the tip of a long, uninsulated delivery tube during a 44° C irrigation. The water that comes out first has been standing in the tube and therefore is much too cool. After 6 sec, water from the reservoir finally appears, but it, too, is initially too cool because part of its heat has been expended in warming up the delivery tube. The water does not reach the proper temperature until the final few seconds of the irrigation. Designers have used two methods to minimize this problem: (1) circulating water from the reservoir through a jacket around the delivery tube, or (2) insulating the deliv-

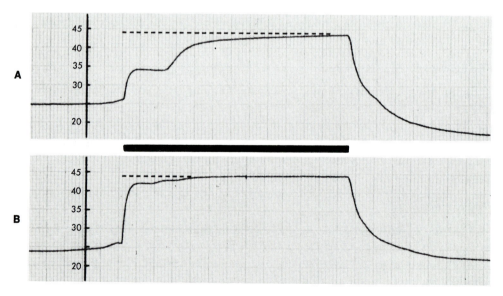

Fig. 3-14. Water temperature at the irrigator tip. **A,** Unjacketed, uninsulated delivery tube. **B,** Jacketed, insulated delivery tube. The horizontal bar indicates the 32-sec irrigation period. The horizontal dotted lines indicate the desired irrigating temperature of 44° C.

ery tube and making it as short as possible. Both methods appear to yield reasonably accurate temperature control of the water. An example of water temperature control by a system that circulates water around the delivery tube is shown in Fig. 3-14, *B*. When using any caloric irrigator, the clinician must experiment in order to find the exact reservoir temperature required to achieve the proper irrigating temperature. For example, it is necessary on one irrigator to maintain the 44° C reservoir at approximately 44.7° C in order to obtain the proper irrigating temperature of 44° C at the tip.

Delivery systems of various caloric irrigators differ in two other significant ways. First, they can operate either by gravity flow or by direct pumping. A direct pumping system should be used, because adequate temperature control of the water at the irrigating tip cannot be obtained in a gravity flow system. The water reservoirs of a gravity flow system must be placed on a shelf high above the patient's head to achieve the proper water flow rate. Therefore, a long delivery tube must be used, and water from the reservoir cannot be circulated through a jacket around it, because the device has no pump.

Second, delivery systems can contain either one delivery tube for both water reservoirs or two tubes, a separate one for each reservoir. The two-tube system is more cumbersome but provides more accurate temperature control. When only one delivery tube is used, it must come to equilibrium at a new temperature each time the technician switches from one reservoir to the other.

The design of the irrigating tip itself is virtually always inadequate; therefore, fabrication of a proper tip is left to the user. Because many patients jerk their heads when the caloric irrigation commences, the irrigating tip should be made of soft rubber so that it will not lacerate the external auditory canal. In addition, the tip should be flared not more than 2 cm from the end so that it cannot be inserted far enough into the canal to touch the tympanic membrane.

The switch that starts the caloric irrigator should be either a button on the irrigating tube itself or a foot switch, enabling the technician to use one hand to hold the delivery tube and the other to hold the catch basin.

No water caloric irrigator currently on the market contains a cooling apparatus; therefore, all of them must be installed where adequate circulation of air is present to keep the ambient temperature below approximately 26° C. Otherwise, it may not be possible to keep the 30° C reservoir cool enough, because it is usually located inside the chassis of the device next to the 44° C reservoir, with no insulation between the two.

Safety features of the caloric irrigator are important considerations. The patient whose ear is being irrigated with water from one electronic device while he is attached via electrodes to another is extremely vulnerable to electric shock. It is therefore imperative that the caloric irrigators meet electrical safety standards as high as those imposed on nystagmographs. The irrigator should also have built-in protection against excessive water temperatures. The temperature of the water in the reservoir is monitored by thermistors, which are notoriously unreliable components. Therefore, backup thermistors should be available to shut off the device if the monitoring thermistors fail, in order to prevent damage to the device or inadvertent irrigation of the ear with water hot enough to scald the patient.

Further considerations are ease of cleaning and filling the reservoir. Irrigators contain complex circuits of water-filled tubes that trap algae and other sediments. If distilled water is used, the amount of contaminant is minimized. However, frequent maintenance is still necessary, and the irrigator should be designed to make maintenance as easy as possible.

Recently a new type of water caloric irrigator was introduced, in which the water continuously recirculates in a closed loop system. The water pressure inflates a rubber balloon placed in the external auditory canal, and the thermal stimulus is transmitted through the walls of the balloon to the temporal bone. Cyr[6] has shown that the closed loop system yields caloric response intensities and reliabilities that are comparable to those produced by the traditional water caloric system.

Air caloric irrigator

Irrigation of the external ear with water has several disadvantages: It can be messy; it exposes the patient to a greater than normal hazard of electric shock; and it is contraindicated in certain cases, such as tympanic membrane perforation or certain other abnormalities of the external or middle ear. These disadvantages are eliminated when the caloric test is performed by irrigating the ear with warm or cool air.

Currently available air caloric irrigators utilize a Peltier thermoelectric device that generates or absorbs heat as a function of current flow through a series of semiconductor elements. As air from an external source (either the standard office air supply or a small air pump) is forced through the thermoelectric unit, it is either warmed or cooled and then delivered to the irrigating tip. Because air loses thermal energy on its journey to the irrigating tip much more readily than water does, the temperature-sensing thermistor of the air caloric irrigator is located near the tip itself.

A considerably greater quantity of air than water is needed to generate the same caloric stimulus, because the transfer of thermal energy between air and tissue is much less efficient than between water and tissue. Capps and others[7] have shown that the intensity of response generated by the standard water caloric stimulus can be generated by an air stimulus if approximately 8 liters of air at either 24°C or 50°C is delivered into the external auditory canal within 60 sec.

Despite the greater safety and convenience of air caloric irrigation, many clinicians are reluctant to accept it, suspecting that it yields less reliable caloric responses. This suspicion was supported by Coats and co-workers,[8] who found that air responses possess significantly greater test-retest variability than water responses. On the other hand, Ford and Stockwell,[9] using great care to place the irrigating tip so that the air stream was aimed directly at the tympanic membrane, found that the reliabilities of air and water responses did not differ. It is probable that placement of the irrigating tip is more critical when using air than when using water.

Air stimulation makes it possible to perform the caloric text in cases where water irrigation is inadvisable, as with tympanic membrane perforation. It should be remembered, however, that the caloric test depends on the assumption that the stimuli to the two ears are equal. If one ear possesses a tympanic membrane perforation or other abnormality that alters the pathway of thermal energy to the labyrinth, the intensity of stimulation of that labyrinth may be quite different from that of the normal labyrinth, even though the caloric stimulus delivered to the two ears is the same.

Paparella and associates[10] have studied caloric responses of patients with a variety of external or middle ear abnormalities and concluded that the presence of a tympanostomy tube or small, dry tympanic membrane perforation does not significantly affect the strength of the air caloric stimulus. However, the presence of a large perforation or an open mastoid or fenestration cavity does affect stimulus strength. In such cases, the assumption of the caloric test would not have been met, and a quantitative comparison between response strengths of the two ears would not be valid. Thus, when air irrigation is used in such cases, the information gained may be of limited value, capable of discriminating whether a response is present or absent, but little else.

Furthermore, Barber and co-workers[11] have shown that bizarre responses can be provoked when an ear with a large tympanic membrane perforation or mastoidectomy cavity is irrigated with warm air. These effects can be attributed, in part at least, to evaporative cooling of the middle ear mucosa; they are more fully described on pp. 180-183.

VISUAL STIMULATORS
Fixation points

An array of points is needed for the patient to fixate upon during calibration of the ENG recording system and the saccade test. These points can be merely an array of spots or lights set at appropriate places on the wall of the laboratory, although several companies manufacture devices that are called "eye movement calibrators." Such devices consist of a frame containing forehead and chin rests to position the patient's head and an array of lights that can be illuminated in various sequences by the technician.

For calibrating horizontal eye movements, one needs a pair of points placed so that its members are separated by the desired distance (usually 20° visual angle) and so that a line between them is horizontal; for calibrating vertical movements, one needs another pair of points separated by the same distance and placed so that a line between them is vertical. The proper distance between members of a pair

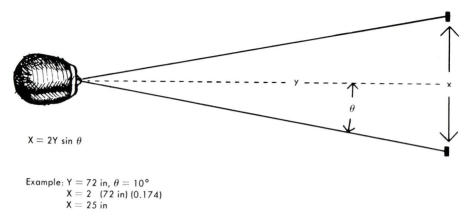

$$X = 2Y \sin \theta$$

Example: Y = 72 in, θ = 10°
        X = 2 (72 in) (0.174)
        X = 25 in

Fig. 3-15. Method of determining the proper distance between fixation points.

depends on the desired visual angle and the distance of the patient from the array; it can be determined by the method shown in Fig. 3-15. One can place another array of points on the ceiling of the laboratory so that recalibrations can be performed between caloric irrigations without returning the patient to the sitting position.

Moving target

A moving visual target is needed for the tracking test. A target can be constructed simply by attaching a brightly colored object to a string and swinging it back and forth. The target should move through approximately 30° visual angle, and its peak velocity should be about 40°/sec. The peak velocity of the target depends on its excursion and on its period of oscillation, which in turn depends on the length of the pendulum. These values can be calculated by the method shown in Fig. 3-16.

Optokinetic stripes

An optokinetic stimulus is needed to perform optokinetic testing. When a neuro-ophthalmologist tests optokinetic responses, he usually passes a striped cloth before the patient's eyes. However, ENG recording of optokinetic nystagmus permits a quantitative estimate of the nystagmus's slow-phase velocity; thus better stimulus control is needed than is provided by a striped cloth. Small motor-driven optokinetic drums that provide a variety of well-controlled stimulus speeds are available. Morissette and associates[12] as well as others have shown that significant pathologic conditions are detected when high target speeds are used; therefore, the optokinetic stimulating device should be capable of generating these speeds. A maximum target speed of 80°/sec should be sufficient. In addition to

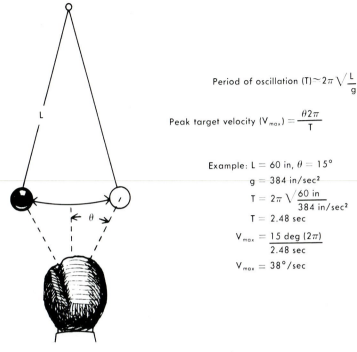

Period of oscillation $(T) \sim 2\pi \sqrt{\dfrac{L}{g}}$

Peak target velocity $(V_{max}) = \dfrac{\theta 2\pi}{T}$

Example: $L = 60$ in, $\theta = 15°$

$g = 384$ in/sec^2

$T = 2\pi \sqrt{\dfrac{60 \text{ in}}{384 \text{ in/sec}^2}}$

$T = 2.48$ sec

$V_{max} = \dfrac{15 \text{ deg } (2\pi)}{2.48 \text{ sec}}$

$V_{max} = 38°/\text{sec}$

Fig. 3-16. Method of determining the peak velocity of a swinging pendulum bob. To determine how far the bob should be displaced to produce the desired excursion, θ, use the method shown in Fig. 3-15.

small optokinetic drums, large rotating cylinders with stripes painted on their inside surfaces are available. When the patient's head is placed inside one of these rotating cylinders, he receives an optokinetic stimulus that covers most of his visual field. Other devices are available that project a large optokinetic stimulus onto a screen or laboratory wall. A device of this type is more convenient than a large rotating cylinder because it is compact and portable, and the stimulus pattern can be changed from horizontal to vertical by turning the device onto its side. It can also be used to project other stimuli besides optokinetic patterns.

ACCESSORY EQUIPMENT AND SUPPLIES
Electrode impedance tester

A device is needed to test the impedance of the recording electrodes. This device can be a battery-powered DC ohmmeter, but an AC impedance meter is preferable. A DC ohmmeter applies a small direct current across the circuit being tested, which is undesirable for several reasons. The current polarizes the electrodes and aggravates the problem of DC fluctuation for at least several minutes after the impedance test. Sometimes it also causes the patient to experience tingling sensations or flashes of light (because the current is applied across the eyes). This sensation is not painful and certainly not harmful to the patient but sometimes creates a loss of rapport. An AC impedance meter avoids both of these problems because it applies an alternating current (usually 10 Hz) across the circuit being tested. Furthermore, the frequency of the alternating test current approximates the frequency content of nystagmus; thus the measured impedance is a more accurate estimate of the impedance present during testing.

Examining table

A patient examining table is also required. The table should allow the examiner to position the patient in both the sitting and supine positions. If this is not feasible, however, a chair can be used to conduct the tests done in the sitting position, and a separate table can be used for tests in the supine position. When the patient is in the supine position, his head should be elevated by 30° for the caloric test. If water is used as the caloric stimulus, a space must be available below the ears for a basin to catch it.

Other equipment

An otoscope is necessary for examining the external auditory canal. A light source, such as an otologist's headlamp, may be used to illuminate the ear during caloric irrigations. A signal generator is desirable for use during maintenance and troubleshooting of the nystagmograph. It should be capable of producing sine, square, and sawtooth waveforms with continuously variable voltages between 0 to 1 v and at continuously variable frequencies between 0.0 to 100 Hz.

The following consumable supplies should be available in the laboratory: electrode paste, adhesive collars (if electrodes with this kind of attaching system are used) or dressing tape, recording paper, spare pens and ink (if ink recording is used) for the nystagmograph, distilled water (if a water caloric irrigator is used), alcohol-soaked or other cleansing pads for cleaning the skin, gauze sponges for abrading the skin, cotton swabs, and spare bulbs and batteries.

REFERENCES

1. Bruner, J. M. R.: Hazards of electrical apparatus, Anesthesiology **28**:396, 1967.
2. Association for the Advancement of Medical Instrumentation: Safe current limits for electromedical apparatus, Arlington, VA, The Association.
3. Geddes, L. A.: Electrodes and the measurement of bioelectric events, New York, 1972, John Wiley & Sons, Inc., pp. 3-43.
4. Geddes, L. A., and Baker, L. E.: Principles of applied biomedical instrumentation, New York, 1968, John Wiley & Sons, Inc., pp. 365-377.
5. Cheng, M., Gannon, R. P., and Outerbridge, J. S.: Frequency content of nystagmus, Aerospace Med. **44**:383, 1973.
6. Cyr, D. G.: An alternative method of caloric stimulation. (In press.)
7. Capps, M. J., Preciado, M. C., and Paparella, M.: Evaluation of the air caloric test as a routine examination procedure, Laryngoscope **83**:1013, 1973.
8. Coats, A. C., Herbert, F., and Atwood, G. R.: The air caloric test: a parametric study, Arch. Otolaryngol. **102**:343, 1976.
9. Ford, C. R., and Stockwell, C. W.: Reliabili-

ties of air and water caloric responses, Arch. Otolaryngol. **104:**380, 1978.

10. Paparella, M. M., Rybak, L., and Meyerhoff, W. L.: Air caloric testing in otitis media (preliminary studies), Laryngoscope **89:**708, 1979.

11. Barber, H. O., Harmand, W. M., and Money, K. E.: Air caloric stimulation with tympanic membrane perforation, Laryngoscope **88:**11-17, 1978.

12. Morissette, Y., Abel, S. M., and Barber, H. O.: Optokinetic nystagmus in otoneurological diagnosis, Can. J. Otolaryngol. **3:**348, 1974.

Chapter 4

Preparations for testing

BEFORE THE PATIENT ARRIVES

Because ENG procedures are almost always conducted by a technician, the preparations for testing are his responsibility. Above all, he must be sure that his instruments meet the electrical safety standards adopted by his hospital. Ordinarily, he will not perform electrical safety checks himself; rather, he will ensure that they are made by a qualified individual at appropriate intervals (usually every 90 days). (Electrical safety is discussed on pp. 43-47). The technician can personally make other checks of his eqiupment and should frequently do so to ensure that it is functioning properly. If using a water caloric irrigator, he should occasionally check the temperature of water that emerges from the caloric irrigator tip with a thermometer, check the timer of the irrigator with a stopwatch, and measure the amount of water delivered during the irrigation. If using an air irrigator, the technician cannot easily measure the air temperature, but he can check the timer. The nystagmograph itself ordinarily requires little maintenance, but the paper supply should be checked. If an ink recording medium is used, ink flow should be inspected. Finally, the technician should be sure that adequate supplies are on hand to conduct the test.

THE PATIENT

The patient wants to know that he is in the hands of someone who is competent and that he has no unpleasant surprises in store for him; therefore, the technician should explain the procedures to the patient in terms that the patient understands before starting the test. He should continue to explain them as he proceeds and answer any questions that the patient has. Some patients would otherwise wildly misinterpret what is being done to them. Some imagine that they are about to receive an injection when the technician begins to clean the skin at the electrode sites with alcohol. Others imagine that the electrodes are being applied for the purpose of giving an electric shock. For most patients, the most frightening part of the procedure is the caloric test. Many of them have come for testing because they have experienced episodes of severe vertigo, and some think that the caloric test will provoke such an attack. The most difficult patient to reassure is one who has previously been subjected to an ice water caloric test that caused vertigo severe enough to provoke nausea and vomiting. The technician can usually reassure this patient by telling him that the test he is about to undergo employs a milder stimulus than the one used in the ice water caloric test.

In trying to soothe the anxious patient, the technician should not mislead him. For example, he should avoid telling the patient that the caloric stimuli will not cause vertigo strong enough to make him nauseated. (It may.) On the other hand, it is foolish for the technician to alarm or alienate the patient needlessly by exaggerating the negative aspects of the test or by deprecating his own competence. Astute management of the patient is a skill gained by experience.

When the patient arrives, the technician should examine him carefully. There are a variety of conditions that will make the technician want to seek instructions from the responsible physician and some that may contraindicate the testing.

Does the patient have an *intracardiac catheter* or *pacemaker with exposed leads?* If so, he is exquisitely sensitive to minute electric currents, and it is unthinkable that he would have been referred for ENG testing. Nevertheless, the technician must not attach electrodes to any patient with suspicious-looking wires that lead under his skin without clarification from the responsible physician.

Each patient who comes for ENG testing will have been examined by an otolaryngologist or neurologist and will have been instructed not to consume any alcoholic beverages and to discontinue certain medications for some interval before the test. However, some patients do not obey. If the technician thinks the patient has been drinking *alcohol,* he should note this fact on the tracing. Recent consumption of alcohol causes nystagmus that might otherwise be interpreted as pathologic (pp. 43-47).

Is the patient taking any *drugs* that would affect the test results? Virtually all patients are taking medications, and some of the drugs can cause nystagmus or other eye movement abnormalities that would otherwise be interpreted as pathologic (p. 149). Many patients do not know the names of the drugs they are taking, nor is the technician expected to identify them or evaluate their effects. Nevertheless, the technician should be aware of the effects of certain commonly used drugs on eye movements, and he should note on the tracing any information on drugs that he elicits from the patient or his hospital chart.

Is the patient *deaf?* If so, communication will be difficult, and testing time will be correspondingly lengthened, although patients with long-standing deafness are often adept at reading lips and gestures. If the patient is wearing a hearing aid, it is helpful to explain the test procedures thoroughly to him before asking him to remove it.

Is the patient *blind?* One cannot obtain ENG recordings from a patient who has nonfunctioning retinas, because these patients have no corneoretinal potential. At best, test results on blind patients will be qualitative because one cannot calibrate the recording system; calibration requires that the patient alternatively fixate on two spots placed on the wall. It is true, however, that a crude calibration can sometimes be obtained for the recently blinded patient by asking him to hold out his two arms with thumbs extended and to "look" alternately back and forth at his two thumbs. One should also beware of the patient with an artificial eye when recording from each eye separately. It is embarrassing to discover that one has been attempting to record from a prosthesis, and these patients are often delighted to know that it was not detected.

Does the patient have *seizures?* The technician should be aware that such a patient might have a grand mal seizure during the examination and should be prepared to deal with it.

Can the patient *freely turn his head?* The technician should be aware that neck rotation can be hazardous for some patients who have cervical disorders, although a physician, not the technician, must bear the responsibility for evaluating this matter.

Other abnormalities may limit the type of testing that can be done, affect the interpretation of the tracing, or contraindicate testing altogether. If the technician encounters an abnormality that he cannot evaluate, he should seek clarification from the responsible physician.

EXAMINATION OF THE EXTERNAL AUDITORY CANAL

The technician should examine both external auditory canals of every patient with an otoscope. He should satisfy himself that both tympanic membranes are intact. If a perforation exists, irrigation of that ear with *water* is contraindicated. Irrigation with *air* is not contraindicated, although the responses that are provoked may be bizarre and difficult to interpret (p. 180). Furthermore, if the perforation is large and the labyrinth is normally responsive, the response is likely to be extremely intense. The technician should not assume the responsibility for detecting perforation; that is the job of the responsible physician. Nevertheless, he should examine the external canal because, after all, he is the one who will perform the irrigations. If the technician is using water as the caloric stimulus and thinks he sees a perforation or any other abnormality that would contraindicate giving the water caloric test, he should seek clarification from the responsible physician before proceeding. Moreover, by examining the external canal, the technician will sometimes detect some other condition that would affect the interpretation of the test. Some types of ear surgery alter the anatomy of the ear and thus change the strength of caloric stimulation that reaches the inner ear. Other naturally occurring anatomical variables, such as size and straightness of the external auditory canal or growth of hair in the canal, also affect the strength of the caloric stimulus. Sometimes an excessive accumulation of cerumen or other debris is present in the canal; this must be removed before the caloric test can be done. If a physician is not available, the technician has no choice but to return the patient to the referring physician for removal of cerumen. However, if a physician is available, the technician may either ask the physician to remove the cerumen with an ear curette or, upon the physician's recommendation, irrigate the canal with water at body temperature. This procedure is performed in every otolaryngologist's office, usually by a nurse. It may be convenient for the ENG technician to learn this procedure himself, but he should do it only under the supervision of a physician, because sometimes a perforation or other abnormality is disclosed that may require immediate treatment. In addition, the technician needs to inspect the external canal following each caloric irrigation, because he can gain valuable information about the adequacy of the irrigation, as discussed on pp. 174-177.

EVALUATION OF EYE MOVEMENTS

Some technicians examine the patient's eye movements before applying the electrodes. The purpose of this is to detect gross disconjugate eye movements, an infrequent occurrence in daily clinical practice. The patient is asked to gaze back and forth between the technician's fingers, one held about 18 inches distant from the other in the horizontal axis. One eye may move more rapidly or fully than the other.

If disconjugate eye movements are seen by this means, the standard electrode placement (Fig. 4-1) is inadequate. Electrodes must be placed to record from each eye separately (Fig. 4-3) in order to display the disconjugate movements in the tracing.

This procedure may give the technician advance information about what his tracing will show. For example, he will know that the tracing is likely to be of poor quality if he sees that the patient blinks or squints excessively. In addition, he will know whether or not the patient has abnormal nystagmus with eyes open.

APPLICATION AND CHECKING OF ELECTRODES

The direction of eye movement detected by a given pair of electrodes depends entirely on where those electrodes are placed on the skin. The electrode placement that is now standard in clinical ENG is shown in Fig. 4-1. Horizontal movements are recorded from two electrodes called the bitemporal pair, one member of the pair being placed on each temple near the outer canthus of the eye. Vertical movements are recorded from two additional electrodes, one placed above and the other below one of the eyes. A fifth electrode, placed on the forehead, is usually attached to the chassis of the instrument and serves as a ground. Note that it

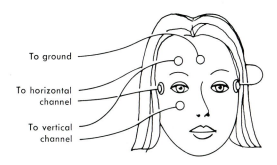

To ground

To horizontal channel

To vertical channel

Fig. 4-1. Standard placement of ENG recording electrodes.

may be inadvisable to attach the patient to ground when using certain types of recording equipment. Check with a biomedical engineer on this point.

The bitemporal electrode pair is maximally sensitive to horizontal eye movements and insensitive to vertical movements; the vertical pair is maximally sensitive to vertical movements and insensitive to horizontal movements. Both pairs are somewhat sensitive to oblique eye movements. To be precise, the sensitivity of a given pair of electrodes to a given eye movement is related to the cosine of the angle between the direction of the movement and a line between the electrodes (Fig. 4-2). Note that one cannot determine the direction of an eye movement from the tracing yielded by one electrode pair alone. To do that, one must compare the directions and strengths of the signals detected by both pairs.

To detect disconjugate eye movements, electrodes must be placed to monitor the movements of each eye separately, as shown in Fig. 4-3; the ground electrode shown may be inadvisable with certain types of recording equipment. Consult a biomedical engineer on this point. Note that a four-channel nystagmograph is needed to make both horizontal and vertical recordings from each eye separately. If only a two- or three-channel nystagmograph is available, the technician must decide which tracings are most important. For example, if he has a two-channel recorder and sees unusual or possible disconjugate eye movements in the horizontal axis, he might wish to record horizontal movements from each eye separately as a supplementary examination, for this purpose giving up recordings of vertical movements. Note that if one eye has restricted excursion, calibration movement will be inaccurate as to amplitude.

The electrodes should be applied as soon as possible after the patient arrives. The longer the electrodes are in place, the better they perform, because the electrical impedance between the electrodes and skin gradually declines as the electrode paste soaks into the skin. The first step is to prepare the skin at the electrode sites by removing skin oils and debris. A variety of methods, such as rubbing the skin with acetone or abrading it with a burr, have been proposed for doing this. However, high-quality tracings can be obtained when the skin is cleaned merely by assiduously rubbing it with an alcohol-soaked pad and abraded by rubbing it with a dry gauze sponge. Preparation of the electrode itself depends on the type of electrode used. Most electrodes in clinical use have a small cup into which the electrode paste is placed. After the paste is applied to this type of electrode, it should be stirred gently with a wooden stick to ensure that it comes into full contact with the Ag/AgCl sensing element. If an air pocket is trapped between the paste and the sensing element, poor electrode performance will be the result. The electrode is then positioned on the skin and held in place with tape. The electrodes must be positioned precisely. The line between members of the horizontal electrode pair(s) should be truly horizontal; the line between members of the vertical pair(s) should be truly vertical; and all lines should intersect the line(s) of sight when the eyes are in the center gaze position.

Once the electrodes are in place, the electrical impedance of all electrode-skin

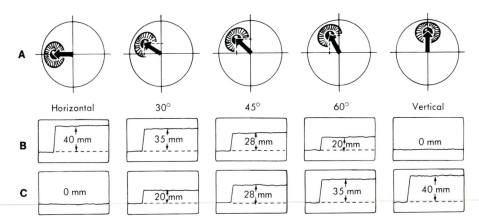

Fig. 4-2. Effect of the direction of a 40° eye movement on the signals recorded by bitemporal and vertical electrode pairs. **A,** Eye movement. **B,** Tracings yielded by bitemporal pair. **C,** Tracings yielded by vertical pair. The indicated pen deflections assume a recording system calibration of $1° = 1$ mm.

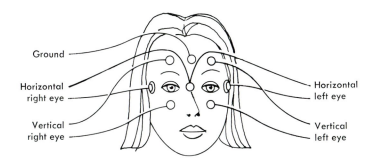

Fig. 4-3. Placement of ENG electrodes to monitor the movements of each eye separately.

junctions should be checked. This is usually done by connecting one terminal of the impedance meter to the lead of the ground electrode and the other terminal to the leads of each of the other electrodes in turn. The impedance across any pair should be as low as possible, preferably below 10,000 ohms. If any impedance greatly exceeds 10,000 ohms, eye movement recordings are likely to be poor. In that case, the offending electrode(s) should probably be removed, the skin cleaned again, and the electrode(s) reapplied.

INSPECTION OF THE TRACING AND INITIAL CALIBRATION

After electrode impedances have been checked, the technician can attach the leads to the nystagmograph, turn the machine on, adjust the gain and balance (if DC recording is used) controls to approximately the correct settings, set the paper speed at 10 mm/sec, and ask the patient to gaze straight ahead. If the electrode application is adequate and the recording system is working properly, the tracing will appear clean and unwavering (Fig. 4-4). If the tracing does not look similar to

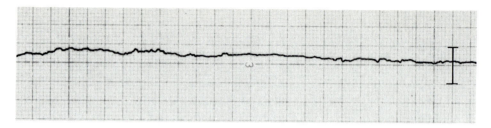

Fig. 4-4. ENG tracing obtained while patient gazes steadily ahead. Bitemporal leads.

the one shown in Fig. 4-4, it contains an artifact that the technician should investigate before proceeding. Various types of artifacts are illustrated and discussed in the next section of this chapter.

If the tracing is satisfactory, the technician should then determine whether the electrode polarities are correct by asking the patient to look alternately right and left, then up and down. The pen(s) of the channel(s) to which the horizontal leads (bitemporal leads if the standard electrode placement is being used or horizontal–right eye and horizontal–left eye if electrodes were placed to record from each eye separately) are connected should move upward when the patient looks rightward and the pen(s) of the channel(s) to which the vertical leads are connected should move upward when the patient looks upward. If the pen on any channel moves in the wrong direction, the appropriate leads should be reversed. This electrode polarity is arbitrary but conventional in clinical ENG.

The next step is to calibrate the recording system. Calibration of the horizontal channel(s) is performed by asking the patient to fixate alternately on two dots (or lights) placed on the wall so that they are separated by a known horizontal distance (usually 20° visual angle). As the patient performs this task, the technician adjusts the gain control of the horizontal channel(s) so that the pen moves 1 mm per degree of eye displacement. Calibration of the vertical channel(s) is done in the same manner while the patient alternately fixates on two other dots placed a known vertical distance apart. Tracings obtained during a calibration are shown in Fig. 4-5. The tracings in Fig. 4-5, *A* and *B,* were obtained using a DC recording system, and the ones in Fig. 4-5, *C* and *D,* were obtained using an AC system. Note that because the AC system contains a low-frequency filter, it causes the pen to drift back toward the baseline as the patient fixates on the target. The rate at which the pen drifts toward the baseline depends on the time constant of the filter; the shorter the time constant, the faster the drift.

Once the calibrations have been performed, the patient is ready for testing. The technician cannot assume, however, that the initial calibration will remain valid for the 'entire testing period. The calibration changes during testing in an unpredictable manner because of fluctuations in the magnitude of the corneoretinal potential;[1] therefore, recalibrations must be performed at regular intervals, preferably before each caloric irrigation and before each of the other tests.

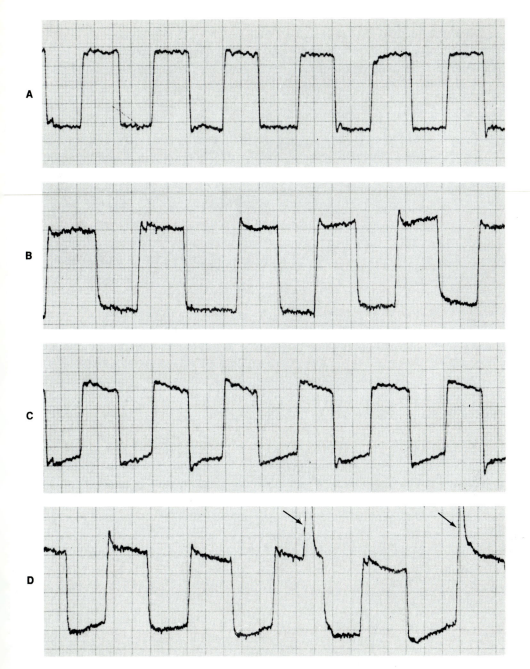

Fig. 4-5. A, Horizontal calibration with bitemporal leads and DC recording system. **B,** Vertical calibration with vertical leads (left eye) and DC recording system. **C,** Horizontal calibration with bitemporal leads and AC recording system (2.5 sec time constant). **D,** Vertical calibration with vertical leads (left eye) and AC recording system (2.5 sec time constant). The arrows denote eye blinks, which may appear prominently in vertical tracings.

ARTIFACTS

When the technician inspects his ENG tracing, he will sometimes find it marred by artifacts that he must investigate and, if possible, eliminate. Two commonly encountered types of artifacts are (1) those caused by electrical interference and (2) those caused by extraneous biologic potentials.

Electrical interference

The source of the most troublesome electrical interference is the half-cell potentials that exist at the electrode-electrolyte interfaces (p. 48). Half-cell potentials are troublesome because they fluctuate and produce a *baseline shift* in the ENG tracing (Fig. 4-6). When changes in half-cell potential are large and persistent, one cannot make DC recordings of eye movements. Unacceptably large fluctuations in half-cell potential usually indicate inadequate electrode application. If the electrodes have been properly applied and the electrode impedance is within acceptable limits, the half-cell potentials should become relatively stable within a few minutes. If they do not, either the electrodes must be reapplied or the recordings must be made using the AC mode with a relatively short time constant (e.g., 3 sec) to keep the recording pen centered in the recording channel.

High electrode impedance also causes a *60-Hz "hum"* to appear in the tracing. It is responsible for the fuzziness of the tracing shown in Fig. 4-6. A large amount of electrical energy at 60 Hz is always present in the ENG laboratory because 60 Hz is the power line frequency. However, under normal circumstances, this signal does not appear in the tracing, because the modern nystagmograph has a high common mode rejection ratio (p. 50) and therefore rejects it. When a 60-Hz hum appears inexplicably in the tracing produced by a recording system known to be functioning properly, it is probably caused by the fact that the impedance of one electrode of a pair is much higher than that of the other. Unequal electrode impedances impair the ability of the recording system to reject common mode electrical signals.

Another source of electrical interference is a *broken electrode wire*. Failure of the commonly used Ag/AgCl electrodes usually occurs because the lead wire breaks at the point where it joins the electrode pellet. As the two broken ends of the wire alternately make and break contact, wild voltage fluctuations occur (Fig. 4-7). One can usually locate the broken wire by jiggling each lead wire in turn at the point where it joins the electrode.

Movement of the electrode relative to the skin is another source of electrical artifact, and it is usually most severe when electrode impedances are high. This type of interference can be easily identified because it appears only when the electrode is touched or pulled upon or when the patient moves the electrode by squinting his eyes or blinking vigorously (Fig. 4-8, *A*). *Movement of the electrode lead wires* can also cause artifact. This artifact is usually a slow-moving potential (Fig. 4-8, *B*). It is easily identifiable because it is present only when the electrode wires are moved.

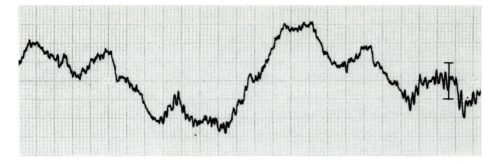

Fig. 4-6. Baseline shift and 60-Hz "hum" usually caused by poor electrode contact.

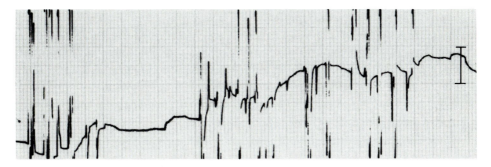

Fig. 4-7. Broken electrode wire.

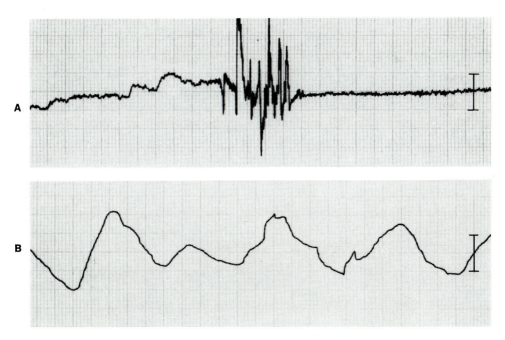

Fig. 4-8. A, Electrode movement. **B,** Electrode lead wire movement.

Electric switches and relays produce *transient voltages* that sometimes appear on the tracing. These are readily identifiable as "blips" (Fig. 4-9).

Extraneous biologic potentials

Several types of potentials of biologic origin cause artifacts in the ENG tracing. The most important and potentially misleading of these biologic potentials arises from *eye blinks*. The technician must recognize eye blinks because the electrical signal they produce can, under certain circumstances, look much like nystagmus. Eye blinks are most prominent on the channel that records vertical eye movements, but they also appear in the horizontal eye movement channel, especially if the line between the electrodes is not exactly horizontal or if the eye blink is vigorous enough to cause movement of the electrodes relative to the skin. Examples of eye blink artifacts are shown in Fig. 4-10; in Fig. 4-10, *A*, the patient is gazing straight ahead with eyes open. In Fig. 4-10, *B*, the patient's eyes are closed. Note that he still produces blinks. Although eye blinks can be misleading, one need never misinterpret them if one recalls that they are always more prominent in the vertical channel than in the horizontal channel. When in doubt, one can ask the patient to blink his eyes several times in rapid succession and note the coincidence between the blinks and the artifacts.

Other biologic potentials are not as important as sources of error, because they are more easily distinguishable. One of these potentials arises from *contraction of facial or neck muscles*. As shown in Fig. 4-11, muscle contraction produces high-frequency electrical signals in the tracing. It can usually be eliminated by saying to the patient, "Relax your face." Another type of potential is caused by the *heartbeat* (Fig. 4-12). It is recognizable as a somewhat distorted EKG signal and is synchronized with the pulse.

MEASUREMENT OF NYSTAGMUS INTENSITY

The primary purpose of ENG is to monitor nystagmus. As we shall see, sometimes the presence or absence of nystagmus is itself significant, but more often the *intensity* of the nystagmus must be known in order to correctly interpret test results. Measurement of nystagmus intensity is usually the task of the ENG technician.

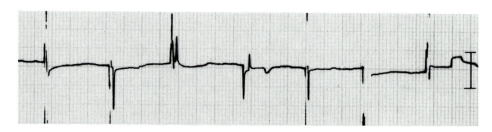

Fig. 4-9. Voltage transients produced by an electrical switch.

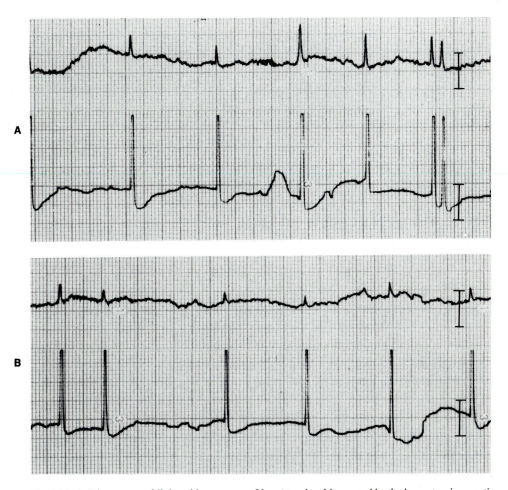

Fig. 4-10. A, Vigorous eye blinks with eyes open. Upper tracing, bitemporal leads; lower tracing, vertical leads. **B,** Same patient, same leads, eyes closed.

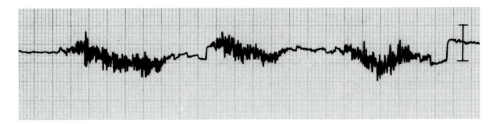

Fig. 4-11. Muscle potentials.

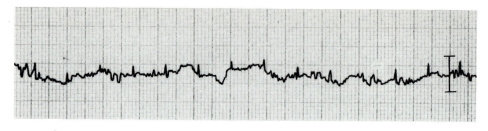

Fig. 4-12. EKG superimposed on tracing.

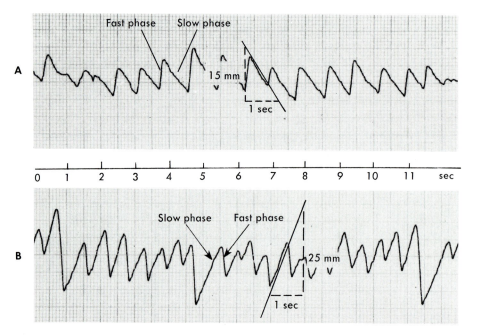

Fig. 4-13. Measurement of nystagmus slow phase velocity. **A,** Right-beating nystagmus. **B,** Left-beating nystagmus. **A** and **B,** The paper moves under the recording pens from right to left, paper speed is 10 mm/sec, and gain is 1 mm/degree of eye rotation.

There are several different types of nystagmus, but measurements of intensity are required only for the types shown in Fig. 4-13. This type of nystagmus consists of alternating fast and slow phases. It appears in the tracing as a sawtooth waveform. In Fig. 4-13, *A,* the nystagmus is right-beating, so designated because the eyes move rightward (upward on the tracing) during the fast phases. They move leftward during the slow phases. In Fig. 4-13, *B,* the nystagmus is left-beating. The eyes move leftward (downward on the tracing) during the fast phases and rightward during the slow phases.

The most widely accepted measure of intensity for this type of nystagmus is *velocity of the eyes during the slow phase*. To make this measurement one must know two values — the speed at which the paper moved under the pens and the gain of the nystagmograph. It is recommended that the paper speed always be set at 10 mm/sec (the paper speed employed in all ENG illustrations in this book) and that the gain always be set so that 1 mm equals 1° of eye movement. Knowing these two values, one can determine how far the eyes have moved (in degrees) during the slow phase within a given amount of time (1 sec), which gives slow phase eye velocity in degrees per second. The slow phase eye velocity of each nystagmus beat is measured separately. Few nystagmus beats last for an entire second, so it is usually convenient to extend the beat by drawing a line parallel to it, as illustrated for one beat in Fig. 4-13, *A,* and one beat in Fig. 4-13, *B*. Move 10 mm horizontally away from the line and then move vertically, counting the millimeters, until the line is intersected again. This procedure showed, for the nystagmus beat displayed in Fig. 4-13, *A,* that the eye moved 15 mm in 1 second. Since 1 mm equals 1° of eye movement, the slow phase velocity of that beat is 15 deg/sec. For the beat shown in Fig. 4-13, *B,* the eye moved 25 mm in 1 second, so the slow phase velocity of that beat is 25 deg/sec. The slow phase velocities of the other beats are measured in the same way.

Where calibration saccade amplitude is found to be other than 1 mm per 1°, calculation of recorded slow phase velocity should be adjusted to read in degrees/sec by the ratio of eye movement in degrees per recorded equivalent in millimeters. Example;

$$\frac{50 \text{ mm}}{1 \text{ sec}} \times \frac{20°}{22 \text{ mm}} = 45°/\text{sec (nearest whole number)}$$

REFERENCE

1. Kolder, H., and Brecher, G.A.: Fast oscillations of the corneoretinal potential in man, Arch. Ophthalmol. **75:**232, 1966.

Chapter 5

Gaze test

NYSTAGMUS

Ocular nystagmus may be defined clinically as a to-and-fro oscillation of the eyes. The eye movement may be a slow-quick jerk (vestibular nystagmus), a movement of approximately equal velocity in opposite directions (pendular nystagmus), or a rotation of the globes (rotational nystagmus). Fig. 5-1 is a schematic representation of pendular and jerk nystagmus. The slow component of jerk nystagmus may be linear, or (in brainstem/cerebellar disease) it may show exponential decrease or increase of velocity.

Pendular eye movements (Fig. 5-1) in the primary position may occur from acquired major visual defect and are common in congenital nystagmus, where they acquire a jerk character on horizontal eccentric gaze. They also occur rarely in acquired CNS disease. Linear jerk nystagmus occurs in peripheral and central vestibular disease, while exponential forms of jerk nystagmus occur only with CNS disease. Rotational nystagmus in the primary position commonly denotes acquired CNS defect but sometimes has congenital origin.

TEST PROCEDURE

The primary purpose of the gaze test is to detect nystagmus with the head in a single position, with eyes in the primary (center gaze) position and deviated horizontally and vertically from this position, and with eyes open and closed.

The gaze test is performed by asking the patient to look straight ahead, then 30° to right, 30° to left, 20 to 30° up, and 30° down. Eye movements are recorded for 15 to 20 sec with eyes open and for about the same time with eyes closed, at least in the primary position (see below). Patients may have difficulty maintaining an eccentric gaze position with eyes closed. It is therefore helpful to ask the patient to hold his arm out with his thumb extended, place his thumb in the proper position, and ask him to "look" at it (by proprioception).

Gaze tests with eyes open are easy for the patient, but with eyes closed the tests are much more difficult. With eyes closed, it may be difficult or impossible to be certain of the eye position, and as nystagmus usually becomes weaker and may disappear entirely with relaxation,[1] the technician must keep the patient alert, a difficult matter in the gaze test. A commonly used method for maintaining alert-

82

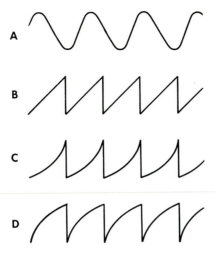

Fig. 5-1. **A,** Pendular nystagmus. **B,** Jerk nystagmus with linear slow phase. **C,** Jerk nystagmus with increasing velocity slow phase. **D,** Jerk nystagmus with decreasing velocity slow phase. (Redrawn from Daroff, R. B.: Nystagmus, Weekly Update: Neurology and Neurosurgery, 1: 1978, Biomedia, Inc., 20 Nassau Street, Princeton, New Jersey, Publishers.)

ness is to present the patient with a series of mental arithmetic (or other) problems, asking him to supply an answer after each question.

When horizontal nystagmus is found with eyes open and *enhanced by eye closure,* the lesion is *peripheral;* nystagmus that is *enhanced with ocular fixation* (eyes open, fixated) and *reduced or abolished by eye closure* results from a *CNS abnormality.* These facts are important to the otoneurologist, and we believe provide justification for retention of search for gaze nystagmus with eyes closed. It is usually easy for the patient to keep the eyes in the primary position with eyes closed, and by this means useful information on the effect on nystagmus of differing conditions of ocular fixation may be obtained. The examiner should probably be very cautious in making diagnostic inferences when the eyes are closed from nystagmus occuring only with gaze deviations.

NORMAL VARIATIONS

Many normal individuals are able to maintain a steady eye position with eyes open or closed in all directions of gaze (Fig. 5-2, *A* and *B*). Others permit their eyes to wander about (Fig. 5-2, *C*), and these deviations can become even larger when their eyes are closed (Fig. 5-2, *D*).

Some normal but highly aroused individuals display *square wave movements* when their eyes are closed (Fig. 5-3, *A*). If square wave movements are prominent when their eyes are open, they may be evidence of a pathologic condition (compare Figs. 5-3, *A,* and 5-23). Normal persons may also exhibit *sinusoidal oscillations* of the eyes at a frequency of approximately 0.3 Hz (Fig. 5-3, *B*). This type of eye movement indicates a person who is drowsy and must be alerted if

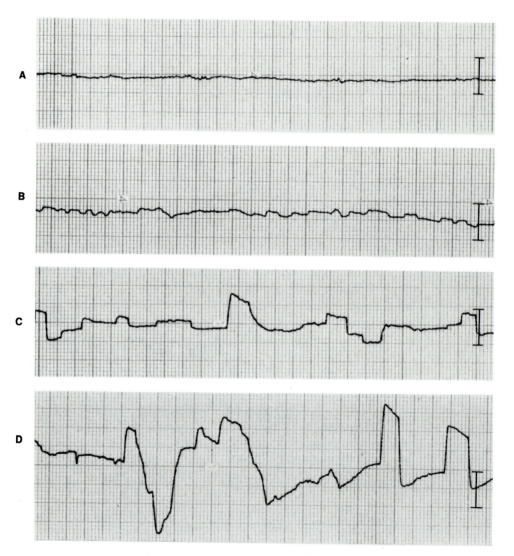

Fig. 5-2. A, Normal individual who maintains steady eye position, eyes open. **B,** Same individual as **A,** eyes closed. **C,** Another normal individual who maintains steady fixation poorly, eyes open. **D,** Same individual as **C,** eyes closed. **A** to **D,** Center gaze, bitemporal leads.

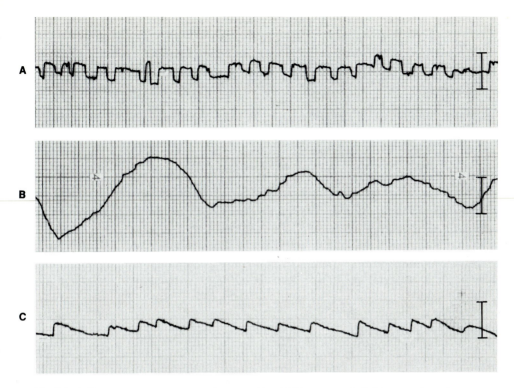

Fig. 5-3. **A,** Square wave movements of a normal but highly aroused individual. **B,** Sinusoidal oscillations of a normal but drowsy individual. **C,** Right-beating nystagmus of a normal individual. **A** to **C,** Eyes closed, center gaze, bitemporal leads.

nystagmus is being sought. If these movements cannot be eliminated by alerting procedures, the patient may be either heavily sedated or possess organic brain damage. Some normal individuals also display *horizontal nystagmus* during gaze testing when their eyes are closed (Fig. 5-3, *C*). It may be present in left, center, or right gaze, but is usually strongest when the person looks in the direction of the fast phase. It is never present when the eyes are open, because it is strongly suppressed by visual fixation. To be considered abnormal, this nystagmus must meet one or more of the criteria listed on p. 143.

JERK NYSTAGMUS
Classification

A. Horizontal
    1. Physiologic (end-point)
    2. Congenital
    3. Pathologic
        a. Peripheral
        b. Central
B. Vertical

Table 5-1. Nystagmus and lesion localization

Nystagmus feature	Peripheral	Central
Axis of movement	Horizontal/horizontal-rotatory	Horizontal, vertical, oblique
Direction	Single	Single, two or more
Waveform slow phase	Linear	Linear, exponential
Eye movements	Always conjugate	Conjugate or disconjugate
Effect of visual fixation	Inhibits	No inhibition

Physiologic. Many normal individuals have a few beats of unsustained gaze-evoked nystagmus in the horizontal axis; these are insignificant. Other normal people show sustained nystagmus with linear slow phase on 30° to 35° lateral gaze deviation; it is always horizontal, bilateral, and fine and may be unequal in the two eyes. According to Daroff,[2] in 40° to 50° lateral gaze positions the slow phase of physiologic nystagmus may be a decreasing velocity exponential.

Congenital. See pp. 102-109.

Pathologic. Virtually every form of jerk nystagmus that is neither physiologic nor congenital is pathologic.

Horizontal nystagmus may result from a peripheral or central lesion. Vertical nystagmus is almost always caused by acquired central disease or intoxication.

Features of nystagmus that usually permit localization of abnormality, peripheral versus central, are given in Table 5-1.

The vestibular system is composed of peripheral pathways (the receptors and afferent nerves), central pathways, and the vestibular nuclei and their connections with the pons, midbrain, and cerebellum.

NYSTAGMUS CAUSED BY PERIPHERAL VESTIBULAR LESIONS

An acute unilateral lesion in the peripheral vestibular pathways causes an asymmetry in the neural input from the two labyrinths (p. 36). The nystagmus that results has characteristics listed in Table 5-1. The nystagmus is unidirectional and, if present in more than a single eye position, is strongest when gaze is in the direction of the quick beat (Fig. 5-4). It is never vertical. Its intensity declines quite rapidly with time, in a matter of hours or days, because of the process of central compensation (adaptation).

Fig. 5-4 is the gaze record of a patient who had had left labyrinthectomy two days earlier. In Fig. 5-4, *A,* the eyes are deviated 30° to the right; nystagmus is obvious with eyes open but enhanced markedly by eye closure. In the primary position (Fig. 5-4, *B*), right-beating nystagmus is present still, though less intense than on gaze to right; on gaze to left (Fig. 5-4, *C*), right-beating nystagmus is faintly detectable, but now less intense in turn than in the primary position. There was no vertical nystagmus. (The absence of nystagmus on gaze to left with eyes closed [Fig. 5-4, *C*] is puzzling; possibly mental alerting was inadequate.)

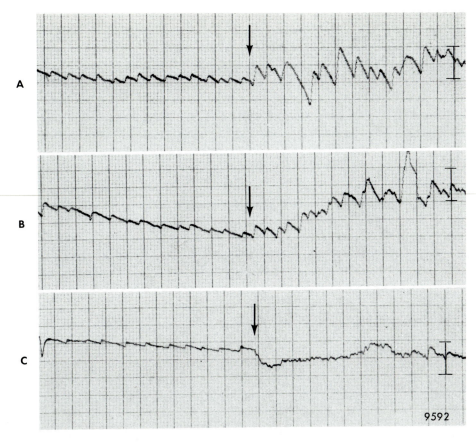

Fig. 5-4. Peripheral vestibular nystagmus 2 days after left labyrinthectomy. A, Gaze to right. B, Center gaze. C, Gaze to left. Nystagmus beats always to right in different eye positions, is enhanced by gaze in direction of the quick beat. Bitemporal leads, eyes open to left of arrows, closed to right. See text.

These features illustrate peripheral localization of lesion. The nystagmus is horizontal, is enhanced by eye closure, is of fixed direction, and varies in intensity with direction of gaze.

Alexander's law, referring to nystagmus in peripheral lesions with eyes open, is illustrated by Fig. 5-4. First-degree nystagmus is said to be present when the nystagmus is found only on lateral gaze in the direction of the quick beat (Fig. 5-4, *A*), second-degree when present in both primary position and lateral gaze in direction of the quick beat (Fig. 5-4, *B*), and third-degree when present also on lateral gaze away from the direction of the quick beat (Fig. 5-4, *C*).

After a peripheral lesion such as labyrinthectomy or severe vestibular neuronitis, the intensity of the nystagmus declines rapidly as the days pass, from CNS compensation. Within a few days of the lesion, third-degree nystagmus has disappeared, followed within a week or ten days by second-degree, and after several months, even the first-degree nystagmus is no longer seen with eyes open. How-

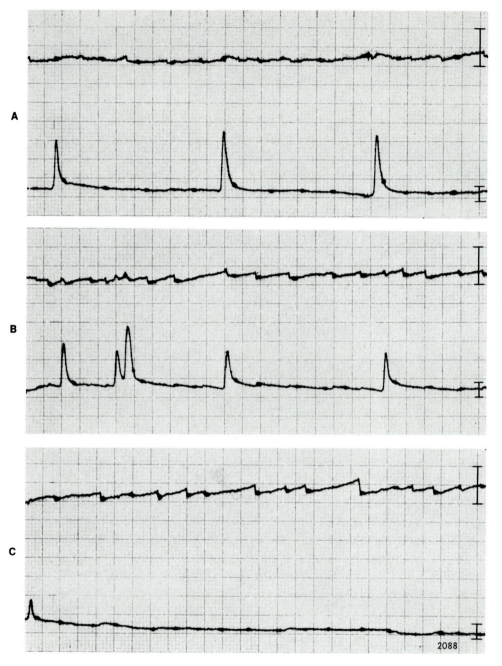

Fig. 5-5. Nystagmus caused by peripheral vestibular lesion, enhanced by upward and downward gaze. **A,** Center gaze. **B,** Upward gaze. **C,** Downward gaze. **A** to **C,** Eyes open; upper tracing, bitemporal leads; lower tracing, vertical leads. The large spikes in the vertical tracings do not represent nystagmus; they are eye blinks.

ever, after labyrinthectomy (or severe lasting vestibular neuronitis), nystagmus beating away from the side of the lesion may sometimes be detected with eyes closed (and effective mental alerting) for many months or even years.

Sometimes horizontal nystagmus that is faint or absent when the eyes are open in the primary position is enhanced by upward gaze, downward gaze, or both. The patient in Fig. 5-5 has right-sided vestibular neuronitis of 5 days' duration. Nystagmus is marginal or absent when the eyes are in the primary position (Fig. 5-5, *A*), but it becomes evident on upward gaze (Fig. 5-5, *B*) and especially on downward gaze (Fig. 5-5, *C*). Note that the nystagmus is horizontal, not vertical. Although horizontal nystagmus on upward gaze is very common in congenital nystagmus (p. 106), it may also be caused by an acquired lesion (Fig. 5-5, *B*). Differentiation from congenital horizontal jerk nystagmus would be sought through identification of other features of congenital nystagmus.

NYSTAGMUS CAUSED BY CNS LESIONS

Nystagmus caused by CNS lesions differs from that caused by peripheral vestibular lesions. CNS nystagmus may be horizontal, vertical, oblique, or rotatory. It may show considerable variation in amplitude. If caused by a fixed, stable lesion, it declines slowly, if at all, with time. It is usually enhanced by ocular fixation. If horizontal, CNS nystagmus is most often bilateral (bidirectional), beating to the right on gaze rightward and to the left on gaze leftward. The term *gaze-evoked nystagmus* refers to CNS nystagmus whose slow phase has a decreasing velocity exponential form (Figs. 5-1 and 5-8); this is also designated as *integrator nystagmus* because modern ocular control theory postulates that it occurs from a defect in the mechanism (integrator) responsible for maintenance of eccentric horizontal gaze, in brainstem or cerebellar system.[3] Central vestibular nystagmus has a linear shape of slow phase.

Bilateral horizontal gaze nystagmus

The most common form of nystagmus of CNS origin is bilateral gaze nystagmus in the horizontal axis. The nystagmus beats to the right on rightward gaze and to the left on leftward gaze. Brainstem cerebellar system localization is the diagnostic meaning of the finding, but the specific cause is quite varied and includes such conditions as compression of posterior fossa structures from a supratentorial mass, drug intoxication or metabolic disturbance, as well as local vascular, neoplastic, or degenerative diseases. It may be accompanied by vertical nystagmus beating upward on gaze upward from the primary position.

Fig. 5-6 illustrates typical features of bilateral gaze nystagmus in a patient with Arnold-Chiari malformation. On gaze 30° to the right, right-beating nystagmus is present when the eyes are open and absent when they are closed (Fig. 5-6, *A*); on gaze 30° to the left, active and high amplitude left-beating nystagmus is seen again only when the eyes are open (Fig. 5-6, *B*).

Another example of bilateral gaze nystagmus is shown in Fig. 5-7. The patient

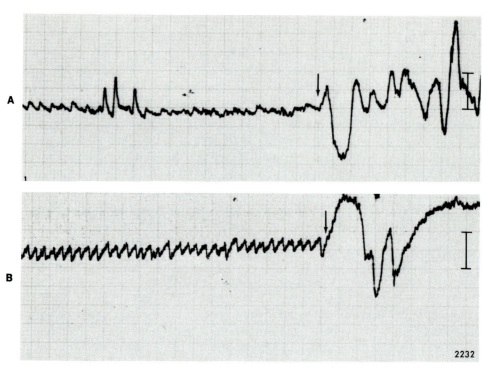

Fig. 5-6. Bilateral (horizontal) gaze nystagmus with eyes open. **A,** Rightward gaze. **B,** Leftward gaze. **A** and **B,** Eyes open to left of arrow and closed to right of arrow, bitemporal leads.

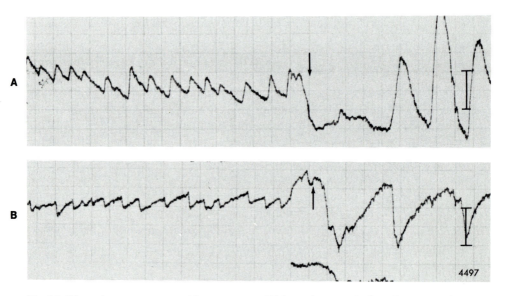

Fig. 5-7. Bilateral gaze nystagmus with eyes open. **A,** Rightward gaze. **B,** Leftward gaze. **A** and **B,** Eyes open to left of arrow and closed to right of arrow, bitemporal leads.

had a large dermoid cyst of the right cerebellopontine angle. High-amplitude gaze nystagmus is present on gaze to the right and left when the eyes are open. There is a suggestion of decreasing velocity exponential shape of the slow component of the nystagmus beat on gaze to right and left, perhaps especially to right.

Large cerebellopontine angle or cerebellar tumors produce characteristic bilateral horizontal gaze nystagmus (Brun's nystagmus).[4,5] On gaze to the side of lesion there is high-amplitude nystagmus with slow beats showing decreasing velocity form; on gaze away from the lesion there is faster small-amplitude nystagmus (see Fig. 5-8). The vestibular nystagmus on gaze away from the lesion may be evoked only with visual inhibition.[4] (See also Fig. 5-9.)

In darkness, normal persons have gaze-evoked nystagmus,[6] slow phases showing decreasing velocity exponential form. It is probable that gaze-evoked nystagmus, present only with eyes closed, also denotes disease sometimes,[7] but if this were an isolated finding in the ENG test battery it would be imprudent to consider it evidence of abnormality. With loss of visual fixation, bilateral horizontal gaze nystagmus recorded with eyes open either becomes absent or shows decreased velocity and frequency, though amplitude usually increases.[8]

Unilateral horizontal gaze nystagmus

The record in Fig. 5-9 is taken from a patient who has right-beating nystagmus that occurs on gaze to the right with eyes open and is absent with eyes closed (Fig. 5-9, *A*); no nystagmus occurs on gaze to the left whether eyes are open or closed (Fig. 5-9, *B*).

This patient is interesting because he has a single lesion that produces effects at two sites within the vestibular system. He is known (on other grounds) to have right-sided acoustic neuroma, a tumor which destroys the afferent vestibular nerve and thus produces by our classification system a peripheral vestibular lesion. The fact that he also has a CNS lesion is demonstrated by Fig. 5-9: only CNS lesions produce gaze nystagmus that is abolished by eye closure. Gaze nystagmus caused by his peripheral vestibular lesion would have been enhanced by eye closure. The CNS damage in this patient was caused by the fact that the acoustic neuroma had grown large enough to compress the brainstem on the right side. It is surprising, and unexplained, that left-beating vestibular nystagmus is absent on gaze to left, even with eyes closed (Fig. 5-9, *B*).

Gaze nystagmus may appear coarse and obvious to the clinician examining the patient's eyes; however, the nystagmus recorded by ENG is often remarkably unobtrusive, featured by low speed and amplitude (Fig. 5-10). Careful scrutiny of the tracing is often required to see it at all. The decision as to whether the abnormality is actually present is facilitated if separate recordings are made from each eye. Nystagmus is recorded stronger in the abducting than adducting eye, and observance of such a tracing is additional assurance of the actual presence of nystagmus. Thus, the left-beating nystagmus shown in the bitemporal leads in Fig. 5-10 is not very impressive, but the certainty of its presence is reinforced by the dual findings of faint nystagmus in the leads for the left eye and none in the right.

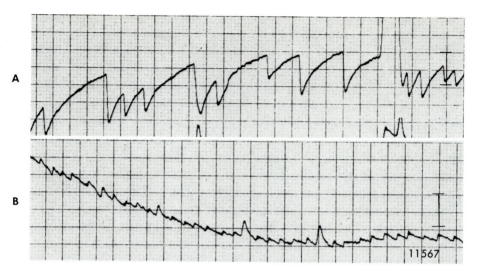

Fig. 5-8. Brun's nystagmus, large acoustic neuroma left side. **A,** Gaze to left; note decreasing velocity exponential form of slow phase, and high amplitude of nystagmus beats. **B,** Gaze to right, low amplitude, higher (than **A**) frequency of beats. Bitemporal leads, eyes open. See text.

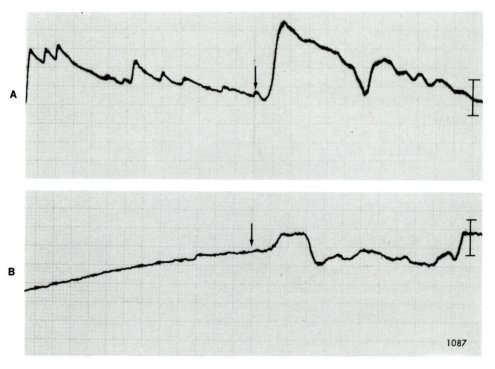

Fig. 5-9. Unilateral gaze nystagmus. **A,** Rightward gaze. **B,** Leftward gaze. **A** and **B,** Eyes open to left of arrow and closed to right of arrow, bitemporal leads. See text.

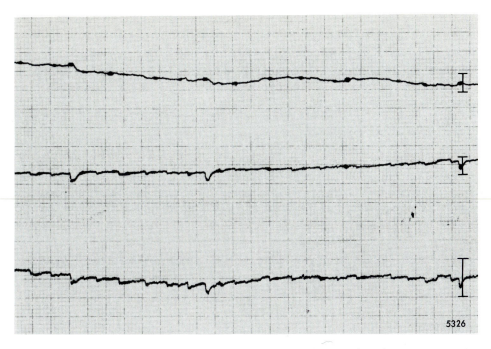

Fig. 5-10. Gaze nystagmus stronger in abducting than in adducting eye. Leftward gaze, eyes open; top tracing, horizontal leads, right eye; middle tracing, horizontal leads, left eye; bottom tracing, bitemporal leads.

Rebound nystagmus

Hood and others[9] have described a form of horizontal gaze nystagmus associated with chronic disease of the cerebellar system. It may be identified on clinical or ENG examination. Hood and his colleagues' description is admirable:

> With eyes in the straight ahead position of gaze, no nystagmus is apparent upon the tracing. With gaze deviation to the right, a brisk nystagmus to the right appears. In the course of some 10 to 20 sec, however, the nystagmus wanes. (It may disappear completely and, on occasions, even reverse in direction. . . .) If at that point . . . , the eyes are returned to the straight ahead position of gaze . . . , nystagmus to the left, not present initially, now makes its appearance. Following the lapse of a further 10 to 20 sec, this nystagmus in turn abates and deviation of the eyes to the left then initiates a brisk nystagmus to the left. This in turn fatigues and a return of the eyes to straight ahead gaze then brings about . . . nystagmus, this time directed towards the right. . . .
>
> The sequence of events described may be repeated indefinitely, provided only that the gaze deviation in each direction is sustained for a sufficiently long period of time, usually of the order of 10 to 20 sec.*

---

*From Hood, J. B., Kayan, A., and Leech, J.: Rebound nystagmus, Brain **96:**507, 1973. By permission of the Oxford University Press.

This clear description is exemplified by the tracing in Fig. 5-11, taken from a patient with olivopontocerebellar degeneration.[10] Not shown in the illustration is the decline of nystagmus with time. In patients with gaze nystagmus it is advisable to monitor lateral deviation of the eyes for a period of up to 20 sec, so as not to overlook this feature, and also to return the gaze to the primary position before gaze to the opposite direction.

Periodic alternating nystagmus

Periodic alternating nystagmus consists of a persistent horizontal or horizontal-rotatory nystagmus that alternates in direction at regular intervals.[11] The usual cycle of periodic alternating nystagmus is about 90 sec in one direction, about 10 sec of absence of nystagmus or irregular downward beats, and 90 sec in the opposite direction. The interval between nystagmus alternations may vary considerably but remains constant for a given patient. At the beginning of a cycle, the nystagmus begins to beat weakly in one direction, builds to a crescendo, gradually declines, and stops. The cycle repeats itself with nystagmus in the opposite direction, then again in the original direction, and so on indefinitely. An example is shown in Fig. 5-12; it was taken from a patient who had vague complaints of dizziness for which there was no definite diagnosis. The patient said that his friends had begun noting his "funny eye movements" 25 years before, when he was in his late twenties.

Periodic alternating nystagmus can be present with eyes open or closed. Oosterveld and Rademakers[12] described twenty-four cases, finding the alternating nystagmus present in seventeen only when the eyes were closed, and in seven with eyes both closed and open. At the middle of the cycle, when nystagmus is strongest, vision is impaired and the patient experiences oscillopsia. Downward beating nystagmus (Fig. 5-15) may coexist.[13] Both forms of nystagmus may occur with pontine and medullary disease, but lower medullary or craniocervical localization of abnormality is frequent in each. Some cases are of congenital origin (as is probable in our case in Fig. 5-12), but many are caused by such acquired lesions as multiple sclerosis, acoustic neuroma, vertebrobasilar ischemia, Arnold-Chiari malformation, cerebellar tumor, or arachnoid cyst.[12-16]

Upbeating nystagmus

Vertical nystagmus may be found on gaze upward or downward or, less frequently, in the primary position. Its recognition is of great clinical importance, because it denotes an acquired lesion, caused by drug intoxication or posterior fossa disease.

Upbeating vertical nystagmus is exemplified in Fig. 5-13. The patient had a large medulloblastoma of the posterior fossa. With eyes open in the primary position, there is extremely faint vertical nystagmus (Fig. 5-13, *A*). On gaze upward, clear upbeating nystagmus is present (Fig. 5-13, *B*).

The gaze records of a patient with Wernicke's encephalopathy (thiamine defi-

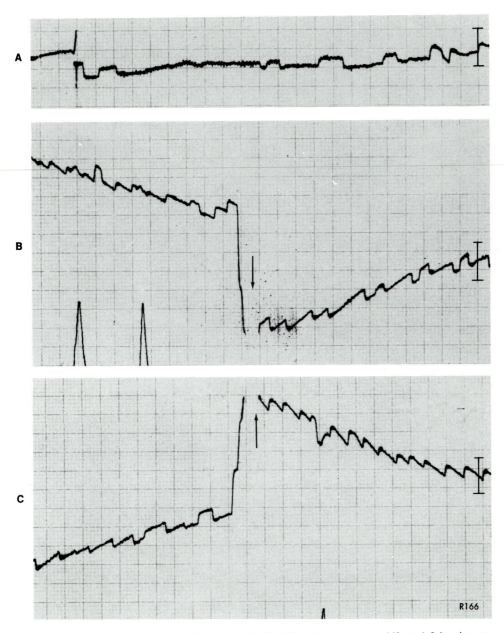

Fig. 5-11. Rebound nystagmus. **A,** Center gaze. **B,** Right-beating nystagmus shifts to left-beating nystagmus when eyes move from rightward to center gaze (arrow). **C,** Left-beating nystagmus shifts to right-beating nystagmus when eyes move from leftward to center gaze (arrow). **A** to **C,** Eyes open, bitemporal leads.

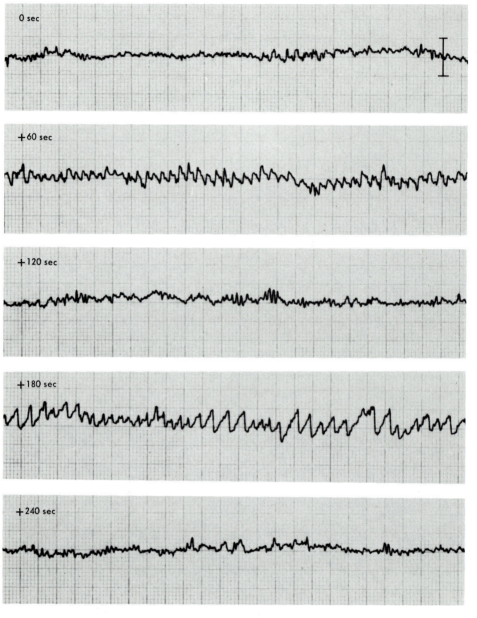

Fig. 5-12. Periodic alternating nystagmus. Successive segments of a tracing recorded during two consecutive cycles. Eyes open and fixating, center gaze, bitemporal leads.

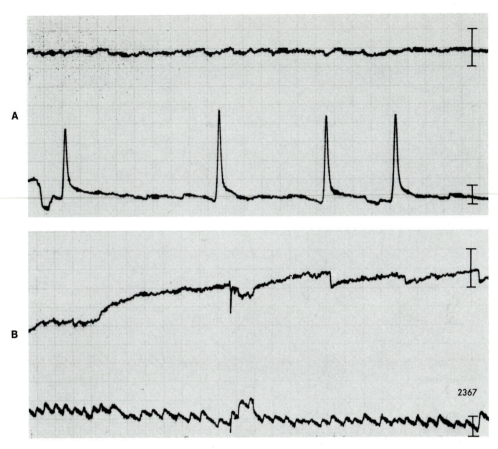

Fig. 5-13. Upbeating nystagmus. A, Primary, or center gaze, position. B, Upward gaze. A and B, Eyes open; upper tracing, bitemporal leads; lower tracing, vertical leads. The large spikes in the vertical tracing in A do not represent nystagmus; they are eye blinks.

ciency in alcoholism) show a variety of interesting features (Fig. 5-14). In the primary position, nystagmus beats obliquely, mainly upward, and a little to the left (Fig. 5-14, A). On gaze to the right, the vector of beat is equally upward and to the right (Fig. 5-14, B); on gaze to the left, the oblique eye movement is mainly directed left, with relatively less vertical component (Fig. 5-14, C). On gaze upward, there is faint, combined vertical downward and horizontal right-beating nystagmus (Fig. 5-14, D), and on gaze downward, the vertical nystagmus is a little more intense than in gaze center (Fig. 5-14, E). Eye closure increases the amplitude of the vertical nystagmus (Fig. 5-14, F).

Daroff[17] describes three types of upbeating nystagmus. The first has large amplitude in the primary position, and this is increased by gaze upward. Lesions of anterior vermis are thought to be responsible. A second type shows low amplitude in the primary position, which, unlike the first, is enhanced by gaze downward.

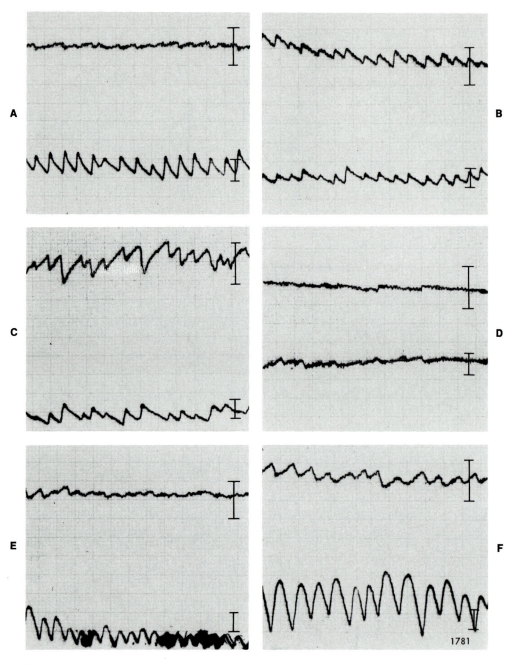

Fig. 5-14. Upbeating nystagmus. **A,** Center gaze. **B,** Rightward gaze. **C,** Leftward gaze. **D,** Upward gaze. **E,** Downward gaze. **A** to **E,** Eyes open; upper tracing, bitemporal leads; lower tracing, vertical leads. **F,** Eyes closed, same leads.

Intrinsic medullary disease is suggested as the cause in this type. A third, intermediate type is similar to the second except for greater amplitude of primary position nystagmus and is said to occur mainly as a manifestation of Wernicke's encephalopathy before treatment (as in our case in Fig. 5-14). However, lesions of the inferior olives in the caudal brainstem have been found in two patients with upbeating nystagmus.[18]

Downbeating nystagmus

Vertical nystagmus beating downward, especially on lateral gaze, is a physical sign that strongly suggests a specific localization, the medullary or medullocervical region. Frequent causes are basilar impression and Arnold-Chiari malformation, which compress the medulla, but any lesion in this specific area may give rise to the finding. Zee and others[19] explain the nystagmus as representing a defect in the transmission of downward velocity information to the neural network that generates smooth pursuit eye movements.

Portions of the gaze record of a patient with lower medullary infarction are shown in Fig. 5-15. Vertical nystagmus is probably absent on gaze in the primary position (Fig. 5-15, *A*); on gaze to the right, there is strong downbeating nystagmus with a horizontal component to the right (Fig. 5-15, *B*); and on gaze to the left, downbeating nystagmus persists, accompanied now by a horizontal component to the left (Fig. 5-15, *C*). Vertical nystagmus is absent on the ENG record on gaze upward (Fig. 5-15, *D*) and (Fig. 5-15, *E*), as well as with eyes closed in lateral gaze (Fig. 5-15, *F*). On another occasion, however, clinical examination of the eyes revealed clear downbeating nystagmus on gaze in the primary position and intensification on gaze downward and changes of head position. Particularly characteristic of this form of nystagmus is its appearance or enhancement on lateral gaze.

Fig. 5-16 shows some of the records of a patient who had recovered from brainstem encephalitis. With eyes in the primary position, there is regular vertical downbeating nystagmus of low speed and a slowed quick component (Fig. 5-16, *A*); on gaze upward, the downbeating nystagmus is a little clearer, though with speed unchanged (Fig. 5-16, *B*); and on gaze downward, it is quite faint (Fig. 5-16, *C*). Thus, downbeating nystagmus here is greater on upward gaze than on downward gaze, unlike the usual pattern in lower medullary disease. On gaze to the right, the vertical nystagmus is of about the same intensity as in the primary position (Fig. 5-16, *D*); but on gaze to the left, it is stronger (Fig. 5-16, *E*).

Other types of nystagmus

Certain types of nystagmus caused by CNS lesions, such as seesaw nystagmus, convergence-retraction nystagmus, and voluntary nystagmus, might also be observed in a gaze test, but patients with such lesions are rarely seen by the otoneurologist. The reader is referred to Daroff,[2,17,20,21] who gives an excellent account of these and other unusual eye movements.

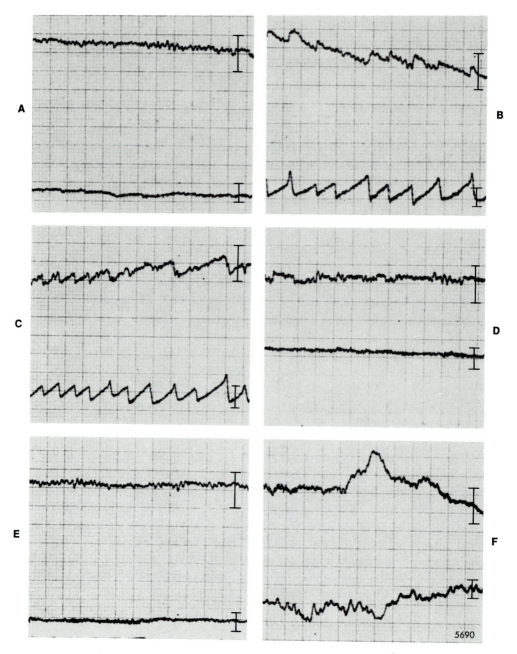

Fig. 5-15. Downbeating nystagmus. **A**, Center gaze. **B**, Rightward gaze. **C**, Leftward gaze. **D**, Upward gaze. **E**, Downward gaze. **A** to **E**, Eyes open; upper tracing, bitemporal leads; lower tracing, vertical leads. **F**, Eyes closed, lateral gaze, same leads.

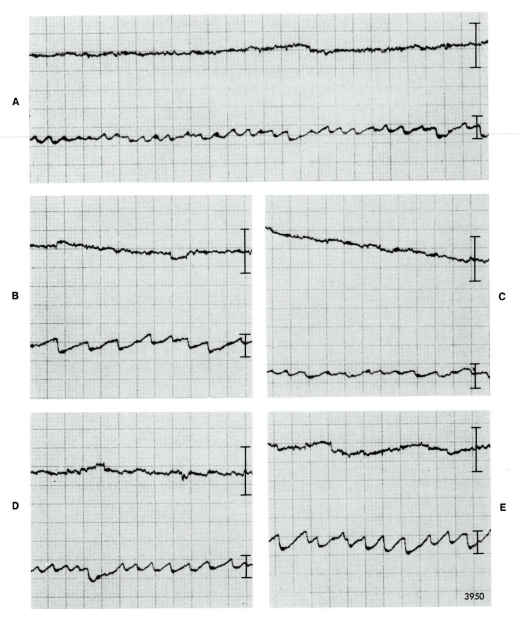

Fig. 5-16. Downbeating nystagmus. **A,** Center gaze. **B,** Upward gaze. **C,** Downward gaze. **D,** Rightward gaze. **E,** Leftward gaze. **A** to **E,** Eyes open; upper tracing, bitemporal leads; lower tracing, vertical leads.

CONGENITAL NYSTAGMUS

The term *congenital nystagmus* is used to describe nystagmus that appears at birth or soon after in an otherwise healthy individual. A good brief account of the entity is given by Gay and others.[22]

Some patients with congenital nystagmus have normal vision, others have visual impairment of varying degree. From an ENG viewpoint, it is extremely important to identify this special form of nystagmus because it results from a fixed, genetic, or developmental brain defect and its recognition makes unnecessary further investigation designed to identify a "cause" for the nystagmus.

Daroff[20] and his colleagues reason that congenital nystagmus "is caused by a high gain instability in the slow eye-movement subsystem," and that "fixation attempt (the effort to see) is the main driving force."

Congenital nystagmus may be pendular or jerk type in character; if pendular in the primary position, it often acquires a jerk-type character on lateral gaze, but the reverse also occurs. Both pendular and jerk-type beats often have an unusual and consistently distorted form that is distinctive for each patient. It may also take other forms; these are distinctive and do not occur in acquired nystagmus.[23]

As shown in Fig. 5-17, on gaze in the primary position, there is jerk nystagmus beating to the left with an irregular, nonlinear slow component, and a spiky termination to both slow and quick beats (Fig. 5-17, *A*). When the same patient gazes to the right, the nystagmus is pendular (Fig. 5-17, *B*); when he gazes to the left, there is intense nystagmus with a curvilinear slow component (Fig. 5-17, *C*).

Fig. 5-18 shows three more examples of congenital nystagmus (each from a different patient). As shown in Fig. 5-18, *A*, the nystagmus is pendular, with sharp spikes at each change of direction, another variety of pendular beat. Pendular nystagmus is highly suggestive of congenital nystagmus or blindness. However, multiple sclerosis is an occasional cause.[24] Fig. 5-18, *B*, shows a similar and rather common deformity of jerk nystagmus. There are spikes at each change of direction; the slow beat is irregular and may be interrupted; and there are a few flattened areas between adjacent beats ("saddling"). A curious type of pendular (actually rotatory) eye movement is shown in Fig. 5-18, *C*. The beats look like regular small triangles. Further examples of the ENG appearance of eye movements in congenital nystagmus are shown by Jung and Kornhuber,[25] and Daroff's line drawings are informative.[26]

One of the main considerations in the diagnosis of congenital nystagmus is its differentiation from gaze nystagmus caused by drug ingestion or acquired lesion. As we have seen, the distorted character of the beat with eyes open may strongly suggest a congenital origin. If congenital nystagmus is suspected, additional features may conveniently be looked for during the gaze test.

In congenital nystagmus, there is a particular direction of gaze called the *null point*, at which the nystagmus declines markedly or stops. The null point is shown in Fig. 5-19. When the eyes are deviated about 30° to left (arrow *A*), left-beating jerk nystagmus is present. As the eyes follow the examiner's target light to the right, the nystagmus declines as it moves to the right of primary position and stops

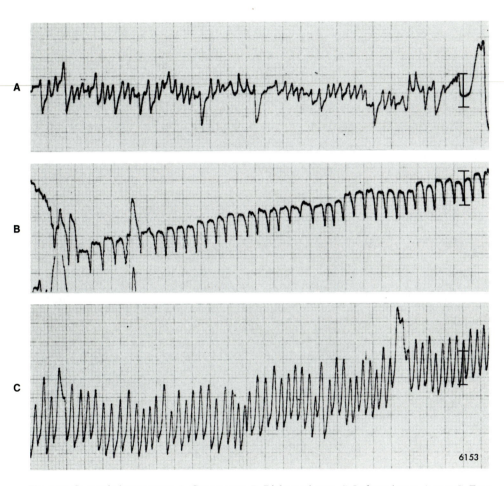

Fig. 5-17. Congenital nystagmus. **A,** Center gaze. **B,** Rightward gaze. **C,** Leftward gaze. A to to C, Eyes open, bitemporal leads. See text.

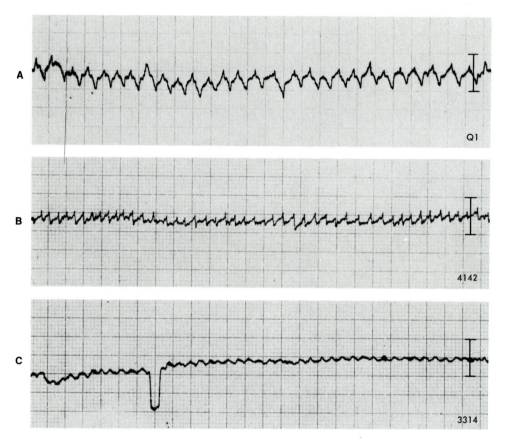

Fig. 5-18. Congenital nystagmus of three different patients. **A** to **C,** Eyes open, center gaze, bitemporal leads. See text.

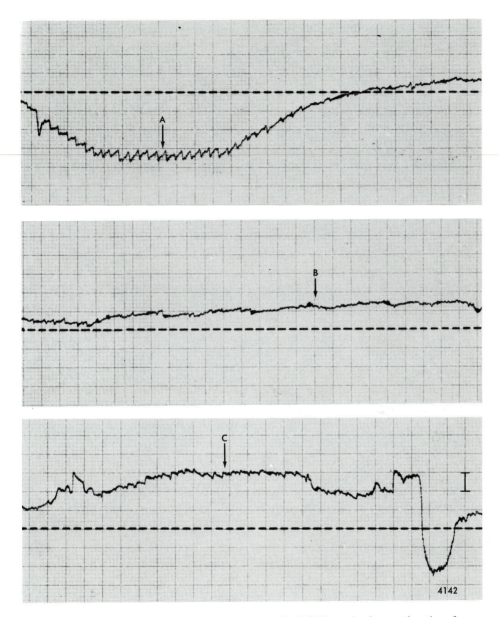

Fig. 5-19. The null point of congenital nystagmus (arrow B). Middle tracing is a continuation of upper tracing, and lower tracing is a continuation of middle tracing. The interrupted horizontal line denotes the primary, or center gaze, position. Eyes open, bitemporal leads. See text.

entirely when it reaches a point 10° to 12° to the right (arrow *B*). With increasing lateral deviation to the right, nystagmus reappears (arrow *C*), now beating to the right. This patient's null point is 10° to 12° to the right of the primary position.

Congenital nystagmus is nearly always horizontal or rotatory, and rarely vertical.[25,27] An important feature of congenital nystagmus is that the *nystagmus on gaze upward is virtually always horizontal, not vertical* (Fig. 5-20). For practical purposes, persistent vertical nystagmus on upward gaze denotes a pathologic or drug-induced condition.

Another important feature of congenital nystagmus is *reduction or abolition of the nystagmus on convergence* (Fig. 5-21). In Fig. 5-21, *A*, the eyes are fixed on a point about 6 ft distant, and the nystagmus is evident. In Fig. 5-21, *B,* when the patient fixates on a point about 2 ft distant, the nystagmus declines markedly and is absent for a few seconds.

For cases in which the differentiation of congenital from acquired nystagmus is difficult, exploration of these three features—null point, upward gaze, and convergence effect—may provide decisive information.

Congenital nystagmus is often reduced or disappears with eye closure, but this is by no means invariable; the same is true of lesional nystagmus of CNS origin. The distortion of beat form in congenital nystagmus is sometimes less marked with eyes closed than with eyes open. When congenital nystagmus occurs behind closed eyes, it may change direction. The record shown in Fig. 5-22 shows that in a single head position the eyes beat first to the left, then after a few seconds in which the nystagmus is absent, to the right. Shifting eye position behind closed lids may account for this occurrence.

Latent (monocular fixation) nystagmus is a form of congenital nystagmus usually associated with strabismus and/or amblyopia. When one eye is occluded, conjugate jerk nystagmus appears, beating toward the unoccluded eye. It may be unilateral or bilateral and, while it may be found frequently in children, it is not a very common occurrence in adult practice.

Fig. 5-23 illustrates the occurrence in a patient on whom no clinical informa-

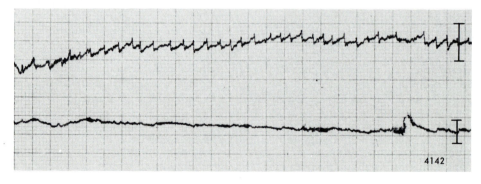

4142

Fig. 5-20. Horizontal congenital nystagmus on upward gaze. Eyes open; upper tracing, bitemporal leads; lower tracing, vertical leads.

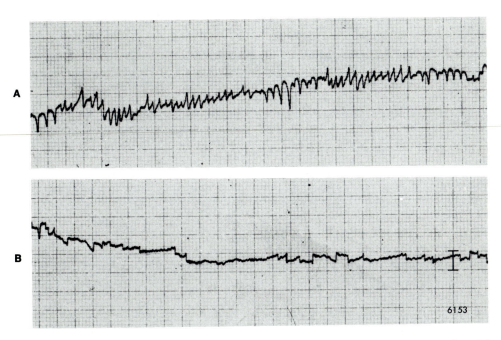

Fig. 5-21. Abolition of congenital nystagmus on convergence. **A,** Fixation on a visual target about 6 ft distant. **B,** Fixation on a target about 2 ft distant. **A** and **B,** Center gaze, bitemporal leads.

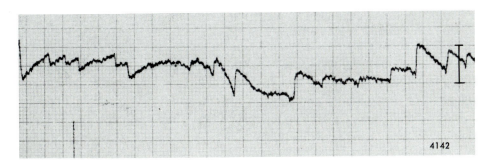

Fig. 5-22. Direction-changing congenital nystagmus with eyes closed. Bitemporal leads.

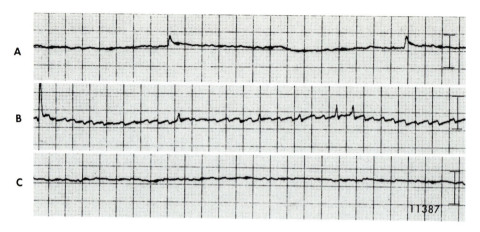

Fig. 5-23. Latent (monocular fixation) nystagmus. **A,** Binocular gaze, eyes open and fixated; nystagmus absent. **B,** Right eye occluded; nystagmus beats to left (though monocular horizontal leads not shown, eye movement is conjugate). **C,** Left eye occluded; nystagmus absent. **A to C,** Center gaze. Bitemporal leads. See text.

tion is available. The eyes are open and in the primary position of gaze in *A, B,* and *C.* In Fig. 5-23, *A,* both eyes are unoccluded; nystagmus is absent. The right eye is occluded in Fig. 5-23, *B,* and left-beating nystagmus of low amplitude is clearly seen. (Only bitemporal leads are illustrated; monocular horizontal leads showed that the nystagmus was conjugate.) No nystagmus is present when the left eye is occluded (Fig. 5-23, *C*).

A variety of pathologic eye movements other than nystagmus may be seen. Most may be regarded as representing fixation instability from cerebellar system, brainstem, or basal ganglia disease.

Square wave jerks are the most common form of fixation instability. They are certainly pathologic when eyes are open and fixated but much more frequent with eyes closed in overalerted, normal individuals (p. 000). They consist of small, conjugate, lateral saccades, succeeded after a 200 msec latency by saccades in the opposite direction which return the eyes to the fixation point. Fig. 5-24 shows regular volleys of square waves that were taken from the same patient (with olivopontocerebellar degeneration), whose rebound nystagmus is illustrated in Fig. 5-11.

Macrosquare wave jerks are large-amplitude (20° or more), intermittent, conjugate, horizontal saccades which move the eyes suddenly off the target, then back again after a latent period off target of only about 80 msec. They occur in dentatorubral (deep cerebellar nuclei and connections) diseases such as multiple sclerosis.[28]

Macrosaccadic oscillations are similar to macrosquare wave jerks but straddle the fixation point. We have not recognized examples of macrosquare wave jerks or macrosaccadic oscillations in our own material, but illustrations of these eye movements are available in the paper of Selhorst and others.[29]

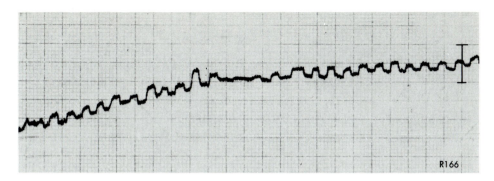

Fig. 5-24. Square wave jerks. Eyes open and fixating, center gaze, bitemporal leads.

Ocular dysmetria is a common sign of cerebellar system disease. (See pp. 40, 116-118.)

Ocular flutter is a momentary rapid, binocular, conjugate, horizontal saccadic oscillation that occurs during gaze in the primary position.[2] It differs from the oscillation that may occur with dysmetria, which always follows a saccadic refixation. In ENG recordings, flutter is triangular in appearance, consisting of several back-to-back saccades[30] (see Fig. 6-10, *B*). Flutter is less common than ocular dysmetria.

Opsoclonus is a bizarre oscillation consisting of rapid, involuntary, chaotic, repetitive, unpredictable, conjugate saccadic eye movements in all directions and persisting during sleep. Daroff[31] suggests the term *saccadomania* as a suitable designation. Flutter and opsoclonus may occur at different times in the same patient, flutter succeeding opsoclonus as clinical recovery progresses. In adults, opsoclonus may reflect the remote effects of a visceral carcinoma. The pathologic abnormality in these patients is usually within or around the dentate nuclei.[30]

Ocular myoclonus is a pendular vertical nystagmus synchronous with oscillation of the palate, larynx, and other midline structures. It is secondary to bilateral pseudohypertrophy of the inferior olivary nuclei and occurs from lesions of the contralateral dentate nucleus or ipsilateral central tegmental tract. The abnormality can be identified only from observation of similar myoclonic movements of palate, etc.

OTHER ABNORMALITIES CAUSED BY CNS LESIONS
Internuclear ophthalmoplegia

Internuclear ophthalmoplegia is caused specifically by a lesion of the medial longitudinal fasciculus in the brainstem between the third and sixth nerve nuclei. Internuclear ophthalmoplegia may be unilateral or bilateral; if unilateral, a common cause is vascular disease, and, if bilateral, demyelination. In fact, bilateral internuclear ophthalmoplegia is strong presumptive evidence of multiple sclerosis.

In early medial longitudinal fasciculus lesions, the only abnormality is saccadic slowing in the adducting eye. A more complete lesion causes nystagmus in the abducting eye, while the adducting eye cannot cross the midline.[32] Slowing or delay of adduction is a consequence of medial rectus muscle weakness from defective innervation and is ipsilateral to the side of the lesion (pp. 17-19).

Fig. 5-25 illustrates these features and is taken from a patient with multiple sclerosis. On gaze to the right, the right eye abducts quickly and fully, while there is obvious slowing and incomplete adduction of the left eye (Fig. 5-25, A). Faint

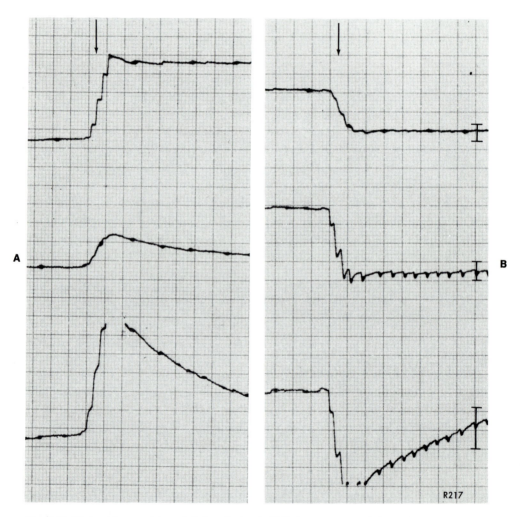

Fig. 5-25. Bilateral internuclear ophthalmoplegia. **A,** Shift from center to rightward gaze (arrow). **B,** Shift from center to leftward gaze (arrow). A and B, Eyes open; top tracing, horizontal leads, right eye; middle tracings, horizontal leads, left eye; bottom tracing, bitemporal leads. See text.

right-beating nystagmus is present in the right eye, but not in the left, though of course it shows in the bitemporal lead. On gaze to the left, the left eye abducts briskly while the right lags and adducts incompletely; monocular (left eye) nystagmus is clear (Fig. 5-25, *B*).

Note that if the bitemporal leads alone had been used, bilateral gaze nystagmus would have been interpreted from the tracing, and the distinctive diagnosis of bilateral medial longitudinal fasciculus lesion would not have been identified.

While we have seen a number of examples of unilateral internuclear ophthalmoplegia, none have sufficiently clear ENG records to warrant illustration. Perhaps the easiest way to identify unilateral internuclear ophthalmoplegia by ENG methods is by recording separately from each eye in the saccade test (p. 119).

ABNORMALITIES CAUSED BY DRUGS

Any surgeon who has examined his patient's eyes in the recovery room soon after general anesthesia is certain to have observed wandering pendular eye movements or jerk (often oblique or vertical) nystagmus. This is a familiar example of drug-induced abnormal eye movements, and many agents are known to cause this effect. Those with particular clinical interest are *barbiturates, phenytoin (Dilantin)* and other anticonvulsants such as *carbamazepine (Tegretol), antihistamines, tranquilizers,* and *alcohol.*

Barbiturates and anticonvulsants, depending on serum levels, may cause a variety of abnormal eye movements. These effects are most likely to be seen in epileptic patients, who often are given one or both of these agents for control of seizures. Sodium pentobarbital (Nembutal) is excreted within 48 hours, but traces of phenobarbital remain in the serum for periods up to 2 weeks from the last dose and could influence the ENG record within that time.

Barbiturates and anticonvulsants act on the CNS, including the reticular formation in the case of barbiturates. Both may cause bilateral gaze nystagmus, as well as saccadic pursuit movements (p. 126). In higher concentration, they are responsible for vertical (usually upbeating) or oblique-upbeating nystagmus, and they can impair or abolish the quick component of caloric nystagmus.

Antihistamines and tranquilizers have a mild sedative effect, and their site of action is presumed to be the CNS. Sekitani and others[33] have shown that diazepam (Valium) in common therapeutic dosage is capable of reducing the firing rate of vestibular nuclei neurons that receive semicircular canal input. While these substances might cause effects similar to those noted for barbiturates and anticonvulsants, we have not observed this in patients who were taking either type of drug alone at the time of testing. However, overall reduction of caloric responses has been found in patients taking 15 to 20 mg of Valium daily at the time of the ENG examination.

The effects of alcohol are similar to those of barbiturates.[34] It causes deterioration of saccadic eye movements and pursuit movements (see Chapter 6), although it is not known to provoke gaze nystagmus, except perhaps in extremely high doses. Alcohol does, however, cause a specific form of positional nystagmus (pp. 149-151).

PITFALLS

There are several possibilities for serious error when one is performing or interpreting the gaze test.

Effects of drugs

Eye movement abnormalities provoked by drugs have just been described. Some drugs, especially barbiturates, antihistamines, and tranquilizers, suppress nystagmus by lowering alertness; therefore, nystagmus might not be observed by the examiner unless he employs alerting procedures assiduously. Other drugs, notably barbiturates (in higher doses) and anticonvulsants, provoke pathologic forms of nystagmus, which the unwary examiner might erroneously attribute to an organic lesion.

All patients referred for ENG examination should be instructed to avoid taking any but life-supporting drugs for 48 hr before testing, and particularly to avoid taking the agents noted here. (The exceptions are epileptic patients, who should not discontinue medications taken to control seizures.) Phenobarbital is not excreted fully within a 48-hr period; however, its use is not very common today. The practice of avoiding drugs for 48 hours has virtually eliminated overt drug effects from our ENG examinations; even when the avoidance instructions have not been carried out, there seems to be remarkably little effect on the records in our patient population. Drug effects would presumably be common in a psychiatric population receiving heavy doses of tranquilizers or in poorly supervised epileptic patients who may develop phenytoin toxicity, but we have had little experience with these circumstances.

It may be impractical to discontinue medications in the case of hospitalized patients. Our strategy, then, is to perform the usual ENG examination, note all medications on the order sheet, and search the records carefully for evidence of drug effect. Any doubts about interpretation should be noted on the report, and, if

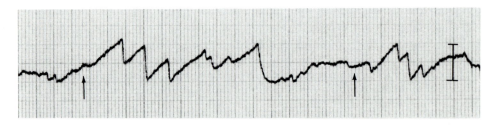

Fig. 5-26. Left-beating nystagmus that appears only when the patient is altered by mental arithmetic (arrows). Eyes closed, center gaze, bitemporal leads.

Table 5-2. Summary of abnormalities observed in the gaze test

Abnormality	Significance	Comment
Horizontal nystagmus (jerk type) Follows Alexander's Law (pp. 87-89)	Acute peripheral vestibular lesion	Always enhanced by eye closure; usually beats toward normal ear
Bilateral gaze nystagmus (pp. 89-91)	CNS lesion	Exclude drugs, congenital nystagmus, excessive eye deviation; usually abolished or reduced by eye closure
Unilateral gaze nystagmus (pp. 91-93)	Peripheral vestibular lesion or CNS lesion	If not enhanced by eye closure, denotes probable CNS lesion
Rebound nystagmus (pp. 93-94)	CNS lesion (cerebellar system)	
Periodic alternating nystagmus (p. 94)	CNS lesion (caudal brainstem)	
Rotatory nystagmus	CNS lesion	Occasionally congenital
Upbeating nystagmus (pp. 94-99)	CNS lesion (anterior vermis, pons, medulla, metabolic)	Exclude drugs
Downbeating nystagmus (p. 99)	CNS lesion (caudal brainstem)	Usually stronger on lateral gaze and may beat obliquely
Pendular nystagmus (p. 102)	Usually congenital or with severe visual loss	Exclude acquired lesion (deep cerebellar nuclei); if congenital may be pendular, jerk type, or both; may appear "distorted," may be abolished by eye closure, remains horizontal on upward gaze, has null point, suppressed by convergence
Square wave jerks (p. 108)	Eyes closed—usually normal, nervous individual Eyes open—CNS lesion (cerebellar system)	
Macrosquare wave jerks (p. 108)	Fixation instability, CNS lesion (cerebellar system)	
Macrosaccadic oscillations (p. 108)		
Ocular flutter (p. 109)		
Opsoclonus (saccadomania) (p. 109)		
Ocular myoclonus (p. 109)	CNS lesion (dentatorubral pathology)	Coincident clonic movement of palate, larynx, etc.
Internuclear ophthalmoplegia (pp. 109-111)	CNS lesion (medial longitudinal fasciculus)	Monocular leads needed to verify; if unilateral, usually denotes vascular disease: if bilateral, usually denotes multiple sclerosis

possible, the examination should be repeated several days after withdrawal of all but life-supporting drugs.

Alertness

Mention has already been made of the fact that CNS-depressant drugs suppress nystagmus by lowering alertness and that patients who are taking these drugs must be pushed hard by alerting procedures, such as mental arithmetic, during the gaze test. Some patients who are not taking drugs must also be constantly alerted, especially during parts of the test in which their eyes are closed. Otherwise, they tend to slip into a state of reverie, and nystagmus that otherwise might be present tends to disappear. Fig. 5-26 illustrates this occurrence. The patient has nystagmus only when he is performing a mental arithmetic problem. When he is unoccupied, his nystagmus is absent.

Vertical nystagmus

During the eyes-closed portion of gaze testing, the vertical channel may show movements that closely resemble nystagmus coincident with horizontal nystagmus. These may result from blink, cross-interference between the horizontal and vertical channels, or actual eye movement (vector movement); the same analytical procedures apply in a gaze record under this circumstance as with caloric testing (pp. 183-186, Figs. 9-18 to 9-20).

REFERENCES

1. Collins, W. E.: Arousal and vestibular habituation. In Kornhuber, H. H., editor: Handbook of sensory physiology, Vol. 6, New York, 1974, Springer Publishing Co., Inc., pp. 361–368.
2. Daroff, R. B.: Nystagmus, weekly update; neurology and neurosurgery, I, Princeton, N. J., 1978, Biomedia, Inc.
3. Daroff, R. B., Troost, B. T., and Dell'Osso, L. F.: Nystagmus and related ocular oscillations. In Glaser, J. S., editor: Neuroophthalmology, New York, 1978, Harper & Row, Publishers, Inc., p. 230.
4. Baloh, R. W., Konrad, H. R., Dirks, D., and others: Cerebellar-pontine angle tumors; results of quantitative vestibulo-ocular testing, Arch. Neurol. 33:507, 1976.
5. Daroff, R. B., and Troost, B. T.: Upbeat nystagmus, J.A.M.A. 225:312, 1973.
6. Daroff, R. B., Troost, B. T., and Dell'Osso, L. F.: Nystagmus and related ocular oscillations. In Glaser, J. S., editor: Neuroophthalmology, New York, 1978, Harper & Row, Publishers, Inc., p. 229.
7. Parker, S. W., and Weiss, A. D.: Some electronystagmographic manifestations of central nervous system disease, Ann. Otol. Rhinol. Laryngol. 85:127, 1976.
8. Baloh, R. W., and Honrubia, V.: Clinical neurophysiology of the vestibular system, Philadelphia, 1979, F.A. Davis Co., p. 115.
9. Hood, J. D., Kayan, A., and Leech, J.: Rebound nystagmus, Brain 96:507, 1973.
10. Brown, J. B.: Diseases of the cerebellum. In Baker, A. B., and Baker, L. H., editors: Clinical neurology, Vol. 2, New York, 1975, Harper & Row, Publishers, Inc., p. 28.
11. Daroff, R. B., Troost, B. T., and Dell'Osso, L. F.: Nystagmus and related ocular oscillations. In Glaser, J. S., editor: Neuroophthalmology, New York, 1978, Harper & Row, Publishers, Inc., p. 232.
12. Oosterveld, M. J., and Rademakers, W. J.: Nystagmus alternans, Acta Otolaryngol. 87:404, 1979.
13. Keane, J. R.: Periodic alternating nystagmus with downward beating nystagmus, Arch. Neurol. 30:399, 1974.
14. Baloh, R. W., Honrubia, V., and Konrad, H. R.: Periodic alternating nystagmus, Brain 99:11, 1976.
15. Daroff, R. B., and Dell'Osso, L. F.: Periodic

alternating nystagmus and the shifting null, Can. J. Otolaryngol. **3:**367, 1974.

16. Towle, P. A., and Romanul, F.: Periodic alternating nystagmus; first pathologically studied case, Neurology **20:**408, 1970.

17. Daroff, R. B.: Ocular oscillations, Ann. Otol. Rhinol. Laryngol. **86:**102, 1977.

18. Baloh, R. W., and Honrubia, V.: Clinical neurophysiology of the vestibular system, Philadelphia, 1979, F.A. Davis Co., p. 116.

19. Zee, D. S., Friendlich, A. R., and Robinson, D. A.: The mechanism of downbeat nystagmus, Arch. Neurol. **30:**227, 1974.

20. Daroff, R. B.: Summary of clinical presentations. In Lennerstrand, G., and Bach-y-Rita, P., editors: Basic mechanisms of ocular motility and their clinical implications, New York, 1975, Pergamon Press, Inc., Publishers, pp. 439–441.

21. Daroff, R. B., Troost, B. T., and Dell'Osso, L. F.: Nystagmus and related ocular oscillations. In Glaser, J. S., editor: Neuro-ophthalmology, New York, 1978, Harper & Row, Publishers, Inc., pp. 236–238.

22. Gay, A. J., Newman, N. M., Keltner, J. L., and Stroud, M. H.: Eye movement disorders, St. Louis, 1974, The C. V. Mosby Co., pp. 66–68.

23. Daroff, R. B., Troost, B. T., and Dell'Osso, L. F.: Nystagmus and related ocular oscillations. In Glaser, J. S., editor: Neuro-ophthalmology, New York, 1978, Harper & Row, Publishers, Inc., p. 223.

24. Aschoff, J. C., Conrad, B., and Kornhuber, H. H.: Acquired pendular nystagmus and oscillopsia in multiple sclerosis; a sign of cerebellar nuclei disease, J. Neurol. Neurosurg. Psychiatry **37:**570, 1974.

25. Jung, R., and Kornhuber, H. H.: Results of electronystagmography in man; the value of optokinetic, vestibular and spontaneous nystagmus for neurologic diagnosis and research. In Bender, M. B., editor: The oculomotor system, New York, 1964, Harper & Row, Publishers, Inc., pp. 435–438.

26. Daroff, R. B., Troost, B. T., and Dell'Osso, L. F.: Nystagmus and related ocular oscillations. In Glaser, J. S., editor: Neuro-ophthalmology, New York, 1978, Harper & Row, Publishers, Inc., pp. 223–227.

27. Walsh, F. B., and Hoyt, W. F.: Clinical neuro-ophthalmology, Vol. 1, Baltimore, 1971, The Williams & Wilkins Co., pp. 279–280.

28. Dell'Osso, L. F., Troost, B. T., and Daroff, R. B.: Macro square wave jerks, Neurology **25:**975, 1975.

29. Selhorst, J. B., and others: Disorders in cerebellar ocular motor control. II. Macrosaccadic oscillation; an oculographic control system and clinico-anatomic analysis, Brain **99:**509, 1976.

30. Daroff, R. B., Troost, B. T., and Dell'Osso, L. F.: Nystagmus and related ocular oscillations. In Glaser, J. S., editor: Neuro-ophthalmology, New York, 1978, Harper & Row, Publishers, Inc., p. 238.

31. Daroff, R. B.: Summary of clinical presentations. In Lennerstrand, G., and Bach-y-Rita, P., editors: Basic mechanisms of ocular motility and their clinical implications, New York, 1975, Pergamon Press, Inc., Publishers, p. 440.

32. Baloh, R. W., Honrubia, V., and Sills, A.: Eye-tracking and optokinetic nystagmus; results of quantitative testing in patients with well-defined nervous system lesions, Ann. Otol. Rhinol. Laryngol. **86:**108, 1977.

33. Sekitani, T., Ryu, J. H., and McCabe, B. F.: Drug effects on the medial vestibular nucleus, Arch. Otolaryngol. **93:**401, 1971.

34. Guedry, F. E., and others: Some effects of alcohol on various aspects of oculomotor control, Aviat. Space Environ. Med. **46:** 1008, 1975.

Chapter 6

Saccade, tracking, and optokinetic tests

SACCADE TEST
Procedure

The saccade test is done during the initial calibration of the recording system (p. 74). As the patient looks back and forth between the two dots on the wall, he performs saccades and yields a tracing that may be inspected for disorders of saccadic eye movement.

Normal variations

When a normal individual makes saccades, his eyes move rapidly[1] and usually stop precisely on each target (Fig. 6-1, *A*). However, some normal individuals consistently undershoot or overshoot the target by a small amount and then must reach it by making one or two small corrective saccades (Fig. 6-1, *B*) or a slow movement, called a glissade[2] (Fig. 6-1, *C*).

Abnormalities

Ocular dysmetria. One function of the cerebellar hemispheres is to control smooth integration of body muscles that function in an agonist-antagonist relationship. Diseases of the cerebellum or its neural connections in the brainstem (collectively known as the cerebellar system) cause defects of limb movement, such as dysdiadochokinesia. The ocular counterpart of dysdiadochokinesia is ocular dysmetria (Figs. 6-2 to 6-4). As shown in Fig. 6-2, when the patient makes a saccade, it is hypermetric, that is, his eyes *overshoot* the target, remain at a point a few degrees beyond the target of 150 to 200 msec, then return to fixate on the target. In this example, overshoot occurs in both directions; however, unidirectional overshoot is more common.

Undershoots of the target, called hypometric saccades, usually carry the same pathologic connotation. Fig. 6-3 illustrates this occurrence in a patient with brainstem-cerebellar findings in association with lung cancer. Another deficit of this type is shown in Fig. 6-4; these tracings indicate both horizontal and vertical eye movements of a man with extensive brainstem infarction involving the cerebellar system and with defective vertical gaze. Horizontal saccades are grossly abnor-

116

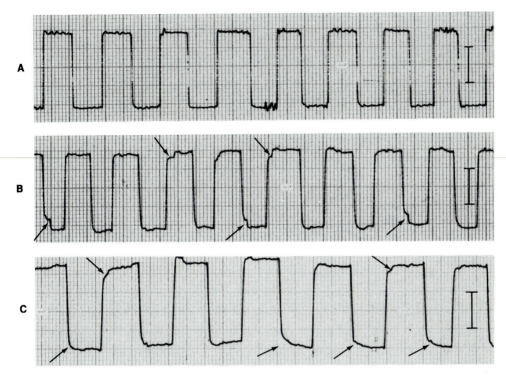

Fig. 6-1. A, Normal saccades. **B,** Normal saccades with occasional corrective saccades (arrows). **C,** Normal saccades with glissades (arrows). **A** to **C,** bitemporal leads.

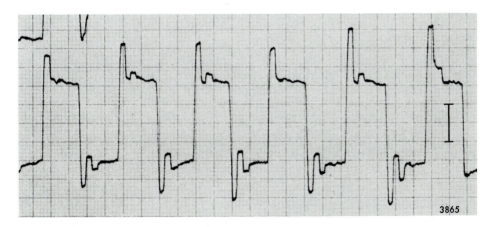

Fig. 6-2. Hypermetric saccades, or overshoots, a form of ocular dysmetria. Bitemporal leads.

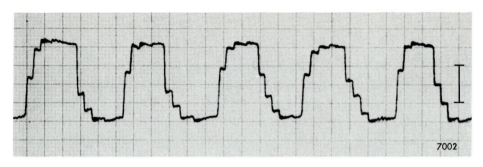

Fig. 6-3. Hypometric saccades, or undershoots, another form of ocular dysmetria. Bitemporal leads.

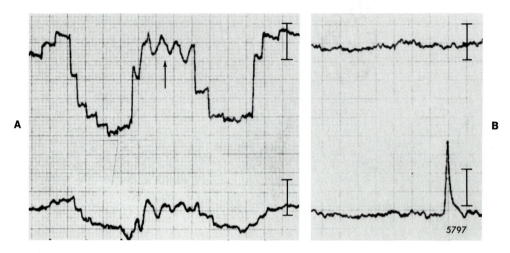

Fig. 6-4. A, Undershoots of the target combined with pendular oscillations while patient is fixating on the rightward target (arrow). **B,** Inability to perform vertical movements. **A** and **B,** Upper tracing, bitemporal leads; lower tracing, vertical leads. See text.

mal (Fig. 6-4, *A*) and are featured by repeated undershoots on attempts to reach the targets to the right and left and a type of jerky, pendular oscillation at 10° ocular deviation to the right. In the attempts at vertical calibration (Fig. 6-4, *B*), no eye movement occurs at all. Note that blinking is retained.

Keep in mind that normal individuals show under- and overshoot to some degree. The finding should be considered pathologic only when it is clear-cut.[3-6]

Saccadic slowing. When a normal individual makes a 20° saccadic movement, his eyes move at a velocity greater than 188°/sec (when measured by a nystagmograph with a 25-Hz low-pass filter).[1] In certain basal ganglia diseases, saccades may be visibly slowed. Fig. 6-5, *A*, shows marked slowing of hosizontal saccades in a patient who has advanced progressive supranuclear palsy.[7-9] This patient also

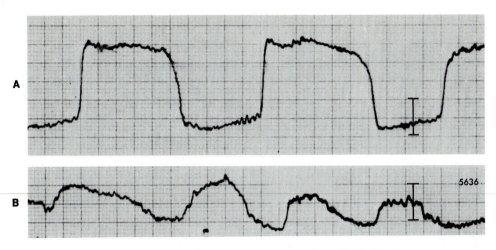

Fig. 6-5. Saccadic slowing. **A,** Bitemporal leads. **B,** Vertical leads.

has nearly complete loss of vertical (especially downward) saccades (Fig. 6-5, *B*), which is even more characteristic of this condition.

Internuclear ophthalmoplegia. In Fig. 6-6, the tracing from the bitemporal leads shows rounding of one side of the upper plateau (small arrows). The tracings of monocular horizontal movements clarify the reason for this occurrence. When an attempt is made to fixate on the target to the right, the right eye moves rapidly to its fixation point (arrow *A*), but the left eye lags behind, attaining fixation about 700 msec later (arrow *B*). The left eye may also have failure of complete adduction and be unable to fully reach the target to the right. In any event, the rounding seen in the tracing from the bitemporal leads is caused by the lag of the adducting eye; the patient has internuclear ophthalmoplegia. Is it unilateral or bilateral? Certainly the lag in adduction of the left eye is obvious, but close scrutiny of the tracings also reveals a slight lag in adduction of the right eye. The internuclear opthalmoplegia is bilateral. Further scrutiny of the tracings reveals another characteristic of the disorder: the patient has nystagmus in the abducting eye, especially when fixating on the target on the right.

Repeated rounding off of the plateaus in the tracing obtained from the bitemporal leads should remind the ENG examiner of internuclear ophthalmoplegia. Separate recordings from each eye should then be made for confirmation.

Pitfalls

Several possibilities for serious error are present when one interprets the saccade test.

Superimposed gaze nystagmus. The clinician should carefully examine the tops and bottoms of the tracing of saccadic movements. Most normal individuals show a smooth regular tracing here, though a minor degree of "quiver," perhaps from

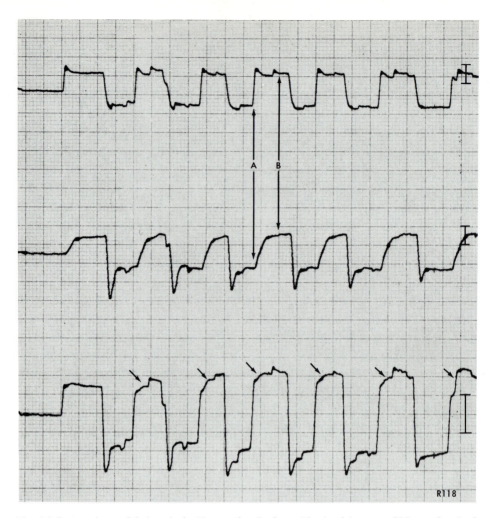

Fig. 6-6. Internuclear ophthalmoplegia. Top tracing, horizontal leads, right eye; middle tracing, horizontal leads, left eye; bottom tracing, bitemporal leads. See text.

muscle potential or microsaccade, is within normal limits. At times bilateral gaze nystagmus will be found on gaze deviations of only 10°. Fig. 6-7 illustrates this occurrence; the patient had medullary compression from Arnold-Chiari malformation, and gaze nystagmus is clearly evident on 10° of angular deviation of the eyes to the right and left (Fig. 6-7, *A*). Vertical gaze nystagmus is also present, especially on upward gaze (Fig. 6-7, *B*).

Superimposed congenital nystagmus. Examples of superimposed congenital nystagmus are shown in Fig. 6-8. Gross pendular eye movements are evident on angular deviations of only 10° to the right and left of the primary position in Fig. 6-8, *A*. These movements are virtually pathognomonic of congenital nystagmus, and

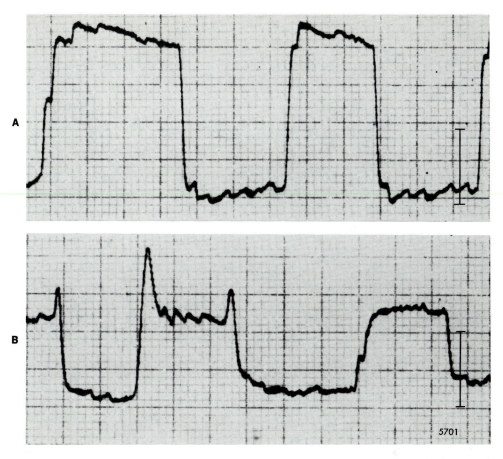

Fig. 6-7. Superimposed gaze nystagmus. **A,** Bitemporal leads. **B,** Vertical leads. The tracing is enlarged (×2) to make the nystagmus more obvious.

the same is true of the distorted jerk-type (rather than pendular) nystagmus found in Fig. 6-8, *B*. The coarse pendular eye movements on gaze to the right and coarse jerks on gaze to the left in the tracings in Fig. 6-8, *C*, can readily be identified as congenital nystagmus. The pattern seen in Fig. 6-8, *D*, also clearly indicates congenital nystagmus; the top plateaus are spiky and irregular. It is not so obvious that the nystagmoid movements at the tops and bottoms of the tracing shown in Fig. 6-8, *E*, are caused by congenital nystagmus; however, on movement to the left, the nystagmus beats to the right, an unlikely occurrence with bilateral gaze nystagmus (compare with Fig. 6-7, *A*).

Congenital nystagmus is easiest to recognize in the gaze test (pp. 82-83).

Drugs. In sufficient dosage, all of the drugs mentioned on p. 111 can produce deterioration of saccadic eye movements. The abnormality most commonly seen is ocular dysmetria. If one detects an abnormality in the saccade test, one must

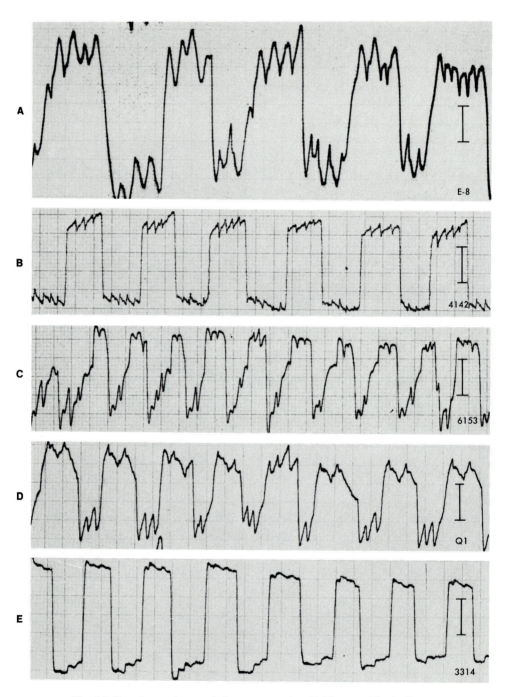

Fig. 6-8. Superimposed congenital nystagmus. A to E, bitemporal leads. See text.

rule out drugs as the cause before accepting the finding as evidence of an organic lesion.

Inattentive patient. A tracing obtained from an inattentive patient is shown in Fig. 6-9. He performs saccades poorly and requires several refixations before he manages to hit the target. This tracing would probably be judged abnormal; thus the technician must coax the patient until he is convinced that he has elicited the patient's best effort. In this case, the patient's poor performance was caused by inattentiveness, not a pathologic condition; normal-appearing saccades were elicited from the same patient after he was coaxed (Fig. 6-9, *B*). In addition to making sure that the patient is performing at his best, the technician must be sure that the patient can see the targets. If the patient has removed his eyeglasses for testing and his saccades are poor, the patient should perform the task again using his eyeglasses.

Eye blinks. The patient who blinks his eyes whenever he performs a saccade can produce a misleading tracing.[10] Sometimes eye blinks produce a tracing that has similarities to one produced when the eyes overshoot the visual target. Fig. 6-10, *A*, shows eye blinks during horizontal saccades, and Fig. 6-10, *B*, shows true saccadic overshoots of a patient with brainstem encephalitis. The tracings obtained from the bitemporal leads have points of similarity, although in Fig. 6-10, *B*, the eyes oscillate back and forth rapidly for three or four excursions after overshoots to the right, from cerebellar system disease. Pathologic overshoots may occur both with and without the postovershoot ocular oscillations. However, the technician can discriminate between blink and overshoot by examining the tracing of vertical movements: blinks produce prominent spikes (Fig. 6-10, *A*), while true overshoots do not (Fig. 6-10, *B*).

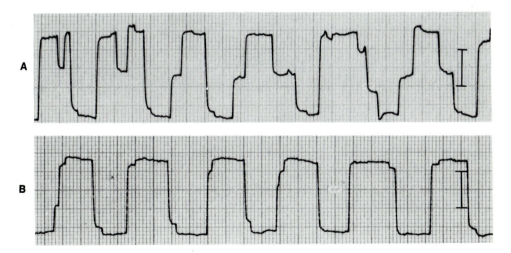

Fig. 6-9. A, Poor saccades from an inattentive patient. **B,** Normal saccades from the same patient after coaxing. **A** and **B,** bitemporal leads.

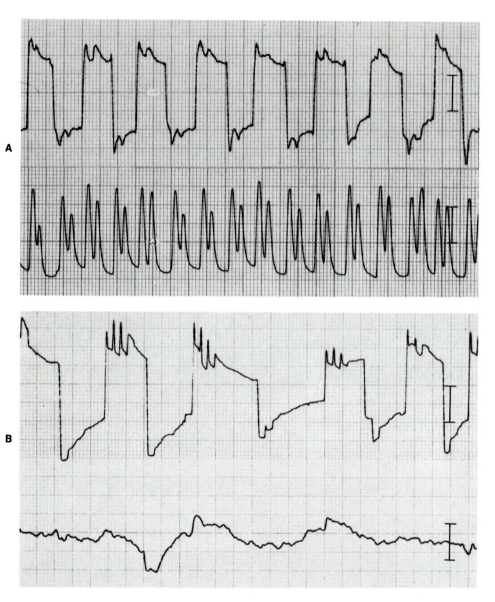

Fig. 6-10. A, Eye blinks during horizontal saccades. **B,** True saccadic overshoots. **A** and **B,** Upper tracing, bitemporal leads; lower tracing, vertical leads.

The tracing in Fig. 6-11 is taken from a patient with cerebellar astrocytoma; it contains both eye blinks and overshoots. The blinks can be easily identified in the tracing of vertical movements; they produce sharp-pointed spikes in the horizontal tracing. In this case, the true overshoots can be distinguished from eye blinks; the overshoots have flattened tops.

Head movement during calibrations. The curious horizontal calibration saccades shown in Fig. 6-12 result from head movement during testing. The eyes move to target, then the head follows and eyes return near to the gaze center position. The technician must steady the patient's head to prevent this movement.

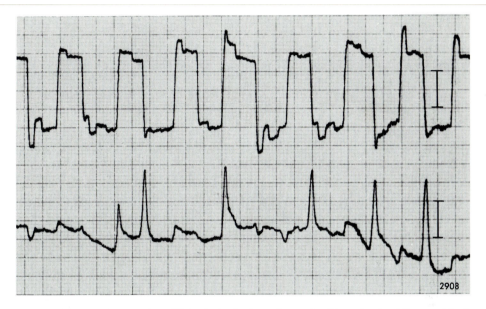

Fig. 6-11. True overshoots and eye blinks during horizontal saccades. Upper tracing, bitemporal leads; lower tracing, vertical leads. See text.

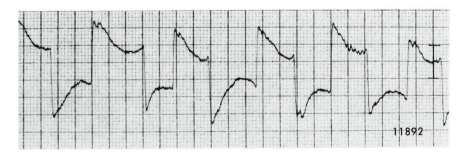

Fig. 6-12. Calibration saccades of normal individual who moves head during testing, bitemporal leads. See text.

TRACKING TEST

Procedure

The tracking test is performed by recording the patient's eye movements while he follows a visual target moving in the horizontal plane. The target may be a brightly colored object suspended on a string and swung back and forth or may be some other arrangement, such as a moving spot of light projected onto a screen. The total excursion of the target should be approximately 30° visual angle, and maximum target speed should not exceed 40° to 50°/sec, because normal individuals begin to have difficulty following targets at higher speeds.[11]

Normal variations

The normal individual is sometimes able to follow the target with negligible error, producing a tracing that is nearly the perfect image of the target motion (Fig. 6-13, *A*). However, some patients follow the target somewhat imprecisely, occasionally allowing it to slip off the fovea and then performing corrective saccades (Fig. 6-13, *B*).

Abnormalities

Saccadic pursuit. When a patient has brainsteam disease involving the pursuit system, he substitutes saccadic movements in varying degree for the smooth tracking capacity. The eye falls behind the moving target briefly, then retrieves it with a saccade. This is the main diagnostic abnormality to identify in the tracing; the term *cogwheeling* is used to describe the appearance of the tracking record when there is marked saccadic pursuit.

Fig. 6-14 shows a comparison of actual eye movement during pendulum tracking and the (superimposed) target movement. The eyes move repeatedly both behind and ahead of the target in saccades and in the interval between attempt to refixate the target.

Fig. 6-15, *A*, illustrates the tracking record of a patient with medulloblastoma of the posterior fossa. There is a moderate degree of saccadic pursuit. The same abnormality is more marked in Fig. 6-15, *B*, taken from a patient with progressive supranuclear palsy. The saccades here are more distinct and well defined than those in Fig. 6-15, *A*, though both records clearly indicate pathologic conditions.

Disorganized and disconjugate pursuit. Fig. 6-16 shows the tracking record of a patient with supranuclear palsy. Two important abnormalities are observed: the first is reduced horizontal gaze capacity, most obvious in the right half of the figure and visible in the tracing from the bitemporal leads; the second is disconjugate eye movement, which requires monocular leads for identification. In the area between the arrows, the right eye first moves slowly to the left, while the left eye remains stationary for about 1 sec; the left eye then moves slowly to the right, while the right drifts aimlessly. Even the tracing from the bitemporal leads gives a clear picture of the wandering, slowed, and inaccurate tracking movement; such a record always indicates a pathologic condition and brainstem localization.

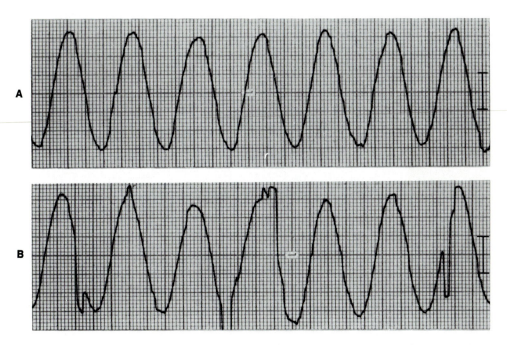

Fig. 6-13. A, Normal tracking. **B,** Less precise but still normal tracking. **A** and **B,** Bitemporal leads.

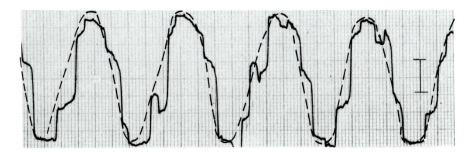

Fig. 6-14. Saccadic pursuit; bitemporal leads; dashed line indicates target motion.

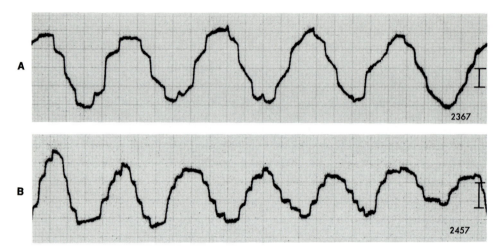

Fig. 6-15. Saccadic pursuit. **A** and **B**, Bitemporal leads.

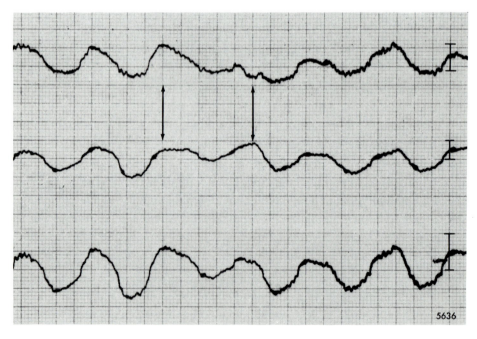

Fig. 6-16. Disorganized and disconjugate pursuit. Top tracing, horizontal leads, right eye; middle tracing, horizontal leads, left eye; bottom tracing, bitemporal leads. See text.

Fig. 6-17 is characterized not only by wavy alteration of the tracking movement, but also by a series of regular rapid overshoots to the right and left, as the eyes repeatedly fail to seize the smoothly moving target. The record is taken from a patient with spinocerebellar degeneration, the same patient whose saccadic movements are shown in Fig. 6-2. The tracing of vertical movements shows that these abrupt movements are not caused by blinks; the single blink on the record is indicated by the arrow.

It is difficult to recognize the tracing shown in Fig. 6-18 as a tracking record, though it is. It has the general appearance of a grossly abnormal calibration record; in fact, this tracing is taken from the same patient as in Fig. 6-4, a man with vertical gaze paralysis and cerebellar system signs from brainstem infarction. Monocular and bitemporal horizontal leads are shown; the first two arrows show brief bursts of disconjugate eye movement, and the third shows not only lack of conjugate movement but also the spiky defect of target fixation, which is clearly identifiable in Fig. 6-4.

Pitfalls

Drugs. The tracking test is like the saccade test in that the patient who is taking certain drugs can register a performance that might be interpreted as abnormal (p. 111). Therefore, drugs must be ruled out as the cause of abnormal tracking before it can be ascribed to an organic lesion.

Noisy record (Fig. 6-19). A noisy record from electronic effect or muscle potential may appear superficially abnormal, though on more careful inspection the nonocular origin of the occurrence becomes evident. Cogwheeling and a noisy

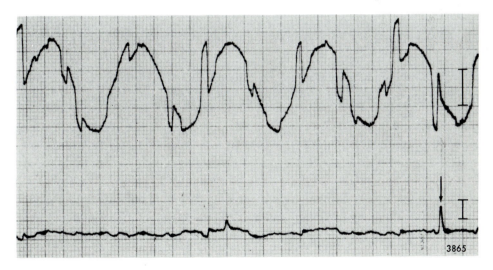

Fig. 6-17. Disorganized pursuit in spinocerebellar degeneration. There are gross overshoots to right and left. Upper tracing, bitemporal leads; lower tracing, vertical leads. Arrow denotes eye blink.

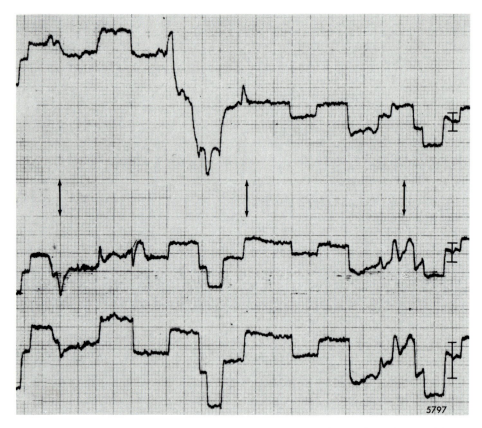

Fig. 6-18. Severely disorganized and disconjugate pursuit. Top tracing, horizontal leads, right eye; middle tracing, horizontal leads, left eye; bottom tracing, bitemporal leads. See text.

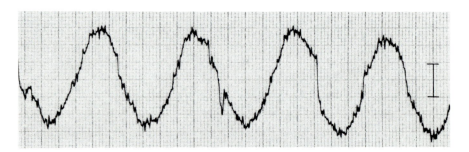

Fig. 6-19. Normal pursuit, noisy tracing; bitemporal leads.

tracing may coexist, of course, as in Fig. 6-20. There appears to be a quite regular 200-msec interval between consecutive saccades in cogwheeling, and this may assist interpretation.

Inattentive patient. An inattentive or uncooperative patient can also yield an abnormal-appearing tracing. Fig. 6-21 is the tracking record of a rather uncooperative and nervous individual who had had a minor head injury and was considered clinically normal. A variety of rapid gaze deviations are seen superimposed on the tracking movement, though there is no suggestion of even minor saccadic alteration of the smooth pursuit when it is present. Moreover, the arrow shows that normal tracking capacity is indeed retained. When tracking performance is poor, the technician should be certain that the patient is performing at his best and that he can see the target.

Head movement. Head movement during tracking causes the irregular excursions shown in Fig. 6-22, taken from the records of a normal individual. The monocular horizontal leads show that the eye movements are conjugate, unlike those shown in Fig. 6-16 (progressive supranuclear palsy). Patterns like this remind the technician to make certain that the patient's head is kept still.

Superimposed gaze nystagmus. If nystagmus is present, the smooth pursuit pattern may be normal (if smooth pursuit brainstem pathways are functioning well),

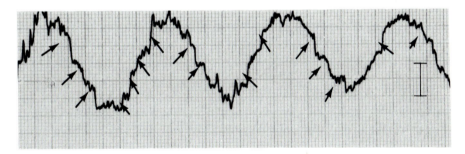

Fig. 6-20. Cogwheeling plus noisy tracing. Arrows indicate saccadic alteration of the pursuit pattern; bitemporal leads.

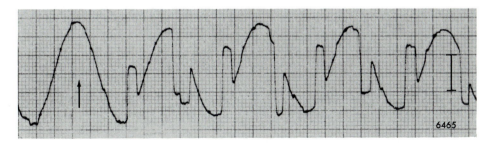

Fig. 6-21. Tracking performance of an inattentive patient. One excursion (arrow) is perfect. Bitemporal leads.

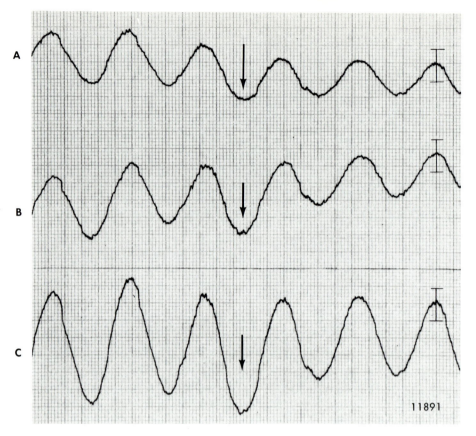

Fig. 6-22. Smooth pursuit record with head movement of normal individual. The excursion of the sinu-
soids is reduced to right of arrows because the patient moves head as well as eyes to follow the target.
Monocular horizontal leads show that eye movements are conjugate—compare with Fig. 6-16. Top
tracing, horizontal lead, right eye; second tracing, horizontal lead, left eye; lower tracing, bitemporal
horizontal leads.

or it may show saccadic pursuit, nystagmus, or both. Fig. 6-23 would be interpreted as showing a minor degree of saccadic pursuit through the middle zones of pursuit movement, with nystagmus at both extremes. The record is taken from a patient with lower brainstem compression caused by Arnold-Chiari malformation.

Fig. 6-24 shows gaze nystagmus to the right and defective pursuit in both
directions, to the right more than to the left. It would be difficult to differentiate
gaze nystagmus from saccadic pursuit from this record alone. Tracings obtained
from the gaze test would have to be studied for clarification. This tracing was tak-
en from a woman with a large right-sided acoustic neuroma; the gaze test revealed
that she had bilateral gaze nystagmus. At surgery, the tumor was found to indent
the brainstem.

Superimposed congenital nystagmus. The tracing shown in Fig. 6-25 was taken
from a patient with congenital nystagmus. The nystagmus beats during eye move-

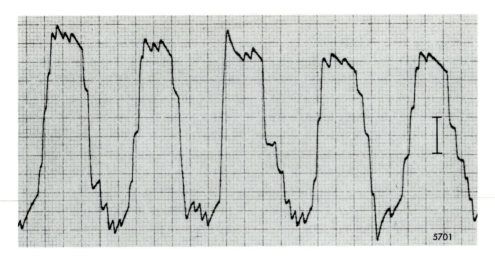

Fig. 6-23. Bilateral gaze nystagmus superimposed on mildly saccadic pursuit movements. Bitemporal leads.

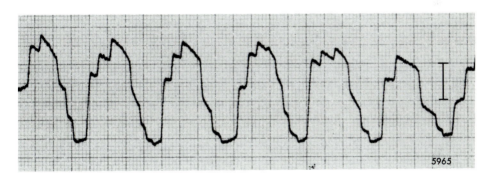

Fig. 6-24. Right gaze nystagmus superimposed on saccadic pursuit movements. Bitemporal leads.

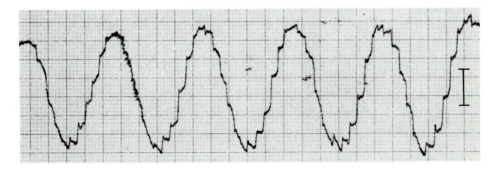

Fig. 6-25. Congenital nystagmus superimposed on pursuit movements. Bitemporal leads.

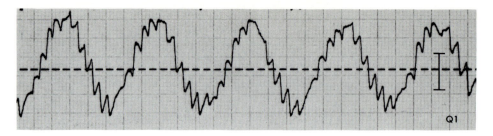

Fig. 6-26. Congenital nystagmus superimposed on pursuit movements. The interrupted horizontal line denotes the primary, or center gaze, position. Bitemporal leads.

ments from left to right are particularly distinct and sharp, many having a rounded character. At both extremes of gaze, there is frank jerk nystagmus, most apparent on gaze to the left. Another example is shown in Fig. 6-26. The jerky eye movements superimposed on the pursuit movements are mainly sharp (some are rounded) and form acute angles with the axis of pursuit movement; it is clear that this is nystagmus rather than cogwheeling (compare with Fig. 6-15).

Congenital nystagmus can usually be distinguished from gaze nystagmus in the tracking record. In congenital nystagmus, nystagmus of some form is seen regularly at or very near the primary position (Fig. 6-26); while in gaze nystagmus, some degree of lateral deviation of the eyes is usually needed for nystagmus production.

OPTOKINETIC TEST
Procedure

The optokinetic test is performed by recording the patient's eye movements as he watches an optokinetic stimulus that is moving horizontally, first to the left and then to the right. A number of stripe velocities should be used; we recommend 10°, 20°, 40°, 60°, and 80° visual angle per sec. If desired, vertical optokinetic responses can be monitored while the patient watches vertically moving stripes.

Normal variations

When a normal individual watches an optokinetic stimulus, the speed of his eyes during the nystagmus slow phase matches the speed of the stimulus, up to a stimulus speed of approximately 30°/sec. As the stimulus speed is increased further, eye speed continues to increase (up to 40° to 50°/sec for stimulus speeds of 80° to 100°/sec), but it falls progressively below target speed. As stimulus speed is increased still further, eye speed declines until the fusion limit is reached.[12]

A normal individual's optokinetic responses are *symmetrical*, which means that for a given stimulus speed, the intensity of his left-beating optokinetic nystagmus (provoked by a rightward-moving stimulus) is approximately the same as that of his right-beating nystagmus (provoked by the leftward-moving stimulus).

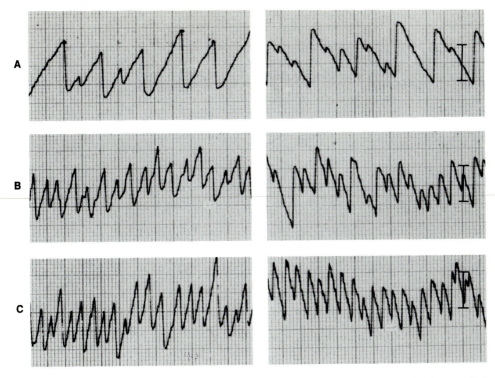

Fig. 6-27. Normal optokinetic nystagmus. **A,** 20°/sec stimulus speed. **B,** 60°/sec stimulus speed. **C,** 100°/sec stimulus speed. **A** to **C,** Left-beating nystagmus is provoked by stripes moving to the right, and right-beating nystagmus is provoked by stripes moving to the left; bitemporal leads.

An example of normal optokinetic nystagmus is shown in Fig. 6-27.

Even when a unilateral peripheral vestibular lesion is present (labyrinthectomy, for example), optokinetic capacity is quite well preserved or only slightly altered; patients with labyrinthectomy may show a transient[13] or permanent minor[12] asymmetry in favor of the normal side.

Abnormalities

Asymmetry. A marked optokinetic asymmetry indicates a CNS abnormality. Our clinical observations indicate that the slow-phase eye speed of left- and right-beating optokinetic nystagmus should differ by about 30°/sec or more, at more than one stimulus speed, in order to be clinically significant.

The following illustrations of optokinetic nystagmus were taken from examinations made with patients facing a flat screen on which vertical stripes were projected with a movie projector at different velocities (Figs. 6-28 to 6-32). Fig. 6-28 is the optokinetic tracing of a patient with medullary compression caused by Arnold-Chiari malformation. At target velocity of 20°/sec, left-beating nystagmus is

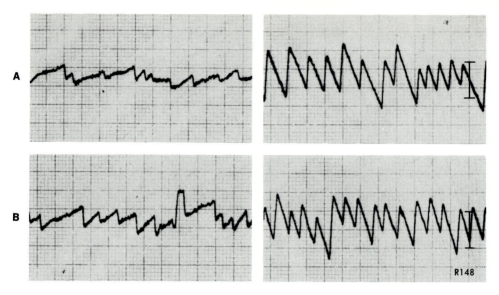

Fig. 6-28. Optokinetic asymmetry. **A,** 20°/sec stimulus speed. **B,** 60°/sec stimulus speed. **A** and **B,** Bitemporal leads.

weaker than right-beating nystagmus (Fig. 6-28, *A*). The asymmetry is even greater at a target velocity of 60°/sec (Fig. 6-28, *B*).

The optokinetic nystagmus responses of a patient with a large glioma of the left temporal lobe are shown in Fig. 6-29. The asymmetry at 20°/sec stimulus speed (Fig. 6-29, *A*) and at 60°/sec stimulus speed (Fig. 6-29, *B*) is marginal. At 120°/sec stimulus speed, however, the asymmetry is marked (Fig. 6-29, *C*). Optokinetic asymmetry, while a definite sign of a pathologic condition, does not localize the abnormality within the central nervous system. Supratentorial lesions of parietal, occipital, and frontal lobes are known to produce asymmetries of direction distinctively related to the side of the lesion.[14] Asymmetry is also found in the patient with brainstem disease, but information on laterality is lacking. For practical purposes, it is sufficient simply to identify significant asymmetry.

"Flat" or declining response intensity to increasing stimulus velocity. If a target is presented at increasing velocities from 20° to 120°/sec, a pattern not characterized by asymmetry but apparently distinctive for brainstem disease may be found.[12] Fig. 6-30 illustrates such a pattern and is taken from a patient with multiple sclerosis. At 20°/sec target velocity (Fig. 6-30, *A*), nystagmus beats to the left and right with equal but low intensity. At 60°/sec target velocity (Fig. 6-30, *B*), nystagmus is present, but the symmetrical low intensity is unchanged from that at 20°/sec stimulus speed. At a still higher target velocity of 100°/sec (Fig. 6-30, *C*), the nystagmus beating to the right and left is still quite symmetrical, but its intensity has, if anything, shown decline from that at the 20°/sec stimulus velocity.

The responses illustrated in Fig. 6-30 are featured by gross failure of nystag-

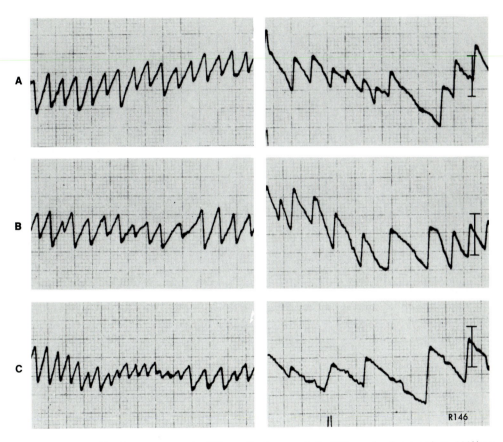

Fig. 6-29. Optokinetic asymmetry. **A,** 20°/sec stimulus speed. **B,** 60°/sec stimulus speed. **C,** 120°/sec stimulus speed. **A** to **C,** Bitemporal leads.

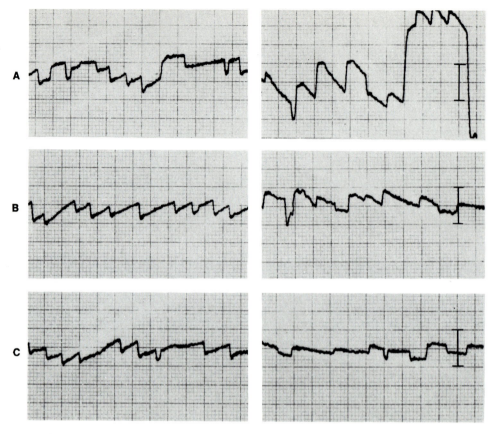

Fig. 6-30. Declining response intensity to increasing stimulus velocity. **A,** 20°/sec stimulus speed. **B,** 60°/sec stimulus speed. **C,** 100°/sec stimulus speed. A to C, Bitemporal leads.

mus intensity to increase with increasing target velocity and by preservation of symmetry of right- and left-beating nystagmus at the different stimulus velocities. If a single stimulus velocity had been used, perhaps 20°/sec, symmetry in response might have led to the erroneous conclusion that optokinetic function was satisfactory.

Certainly this error would be possible from consideration of response to the 20°/sec stimulus shown in Fig. 6-31. This tracing was taken from another multiple sclerotic patient who has a more active response at 20°/sec (Fig. 6-31, *A*) but who has a decline of nystagmus intensity with increasing target velocities of 60°/sec (Fig. 6-31, *B*) and 80°/sec (Fig. 6-31, *C*).

Inversion. The presence of congenital nystagmus may cause marked alteration of the optokinetic record. In Fig. 6-32, *A,* the nystagmus jerks are distorted but beat appropriately to the left when the stripes are moving to the right. However, when the target reverses direction and moves to the left, the nystagmus also beats

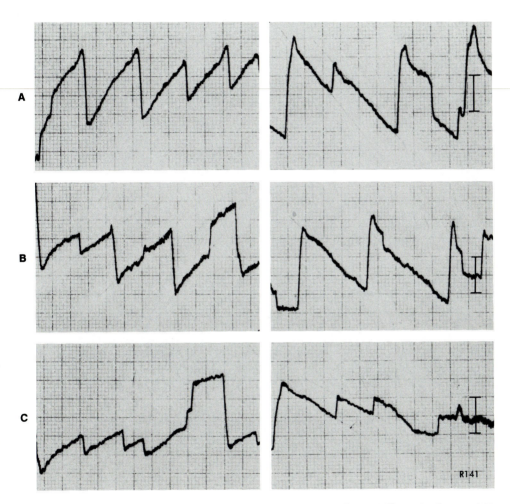

Fig. 6-31. Declining response intensity to increasing stimulus velocity. **A,** 20°/sec stimulus speed. **B,** 60°/sec stimulus speed. **C,** 80°/sec stimulus speed. **A** to **C,** Bitemporal leads.

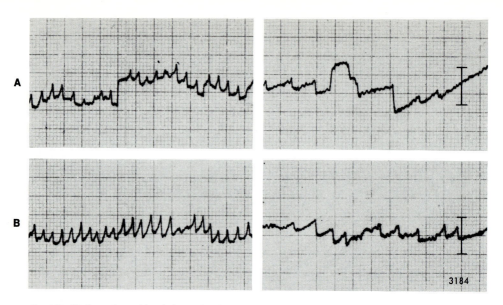

Fig. 6-32. Unilateral optokinetic inversion in a patient with congenital nystagmus. **A,** 20°/sec stimulus speed. **B,** 60°/sec stimulus speed. **A** and **B,** In the left tracing, the stimulus is moving rightward, and in the right tracing, the stimulus is moving leftward; bitemporal leads.

Table 6-1. Summary of abnormalities observed in the saccade, tracking, and optokinetic tests

Abnormality	Significance	Comment
Saccade test		
Ocular dysmetria (pp. 116-118)	CNS lesion (cerebellar system)	Rule out drugs, inattentive or uncooperative patient, poor vision, eye blinks, superimposed nystagmus, head movement
Saccadic slowing (pp. 118-119)	CNS lesion	Rule out internuclear ophthalmoplegia
Internuclear ophthalmoplegia (p. 119)	CNS lesion (medial longitudinal fasciculus)	Monocular leads needed to verify
Tracking test		
Saccadic pursuit (p. 126)	CNS lesion	Rule out drugs, inattentive or uncooperative patient, poor vision, eye blinks, superimposed nystagmus, head movement
Disorganized pursuit (pp. 126-129)		
Disconjugate pursuit (pp. 126-129)	CNS lesion	Monocular leads needed to verify
Optokinetic test		
Asymmetry (pp. 135-136)	CNS lesion	Rule out drugs, inattentive or uncooperative patient, poor vision, congenital nystagmus
Declining response intensity to increasing stimulus velocity (pp. 136-138)	CNS lesion (brainstem)	
Inversion (pp. 138-141)	Congenital nystagmus	May be unilateral or bilateral

to the left, that is, in a direction inappropriate to the stimulus. At increased target velocity (Fig. 6-32, *B*), the same optokinetic pattern is seen. Again the nystagmus beats in the inappropriate direction when the stripes move to the left. The inversion may be bilateral,[15] though we have no record of such an occurrence.

Pitfalls

Drugs and inattentiveness. As with the saccade and tracking tests, the patient who is taking drugs mentioned on p. 111 can produce an abnormal appearing tracing in the optokinetic test. The technician should also be aware that the patient can suppress his optokinetic nystagmus at will by allowing the stimulus to blur before his eyes. Therefore, if nystagmus is of poor quality or is absent in either or both directions, the technician must constantly encourage (indeed, sometimes bully) the patient to watch the stripes and accept the result as valid only when he is sure that the patient is performing at his best.

REFERENCES

1. Boghen, D., and others: Velocity characteristics of normal human saccades, Invest. Ophthalmol. **13**:619, 1974.
2. Weber, R. B., and Daroff, R. B.: Corrective movements following refixation saccades: type and control system analysis, Vision Res. **12**:467, 1972.
3. Weber, R. B., and Daroff, R. B.: The metrics of horizontal saccadic eye movements in normal humans, Vision Res. **11**:921, 1971.
4. Troost, B. T., Weber, R. B., and Daroff, R. B.: Hypometric saccades, Am. J. Ophthalmol. **78**:1002, 1974.
5. Baloh, R. W., and Hourubia, V.: Clinical neurophysiology of the vestibular system, Philadelphia, 1979, F. A. Davis Co., p. 75.
6. Bahill, T. A., and Troost, B. T.: Types of saccadic eye movements, Neurology **29:** 1150, 1979.
7. Steele, J. F., Richardson, J. C., and Olszewski, J.: Progressive supranuclear palsy, Arch. Neurol. **10:**333, 1964.
8. Behrman, S., and others: Progressive supranuclear palsy, Brain **92**:663, 1969.
9. Dix, M. R., Harrison, M. J. C., and Lewis, P. G.: Progressive supranuclear palsy (the Steele-Richardson-Olszewski syndrome), J. Neurol. Sci. **13**:237, 1971.
10. Barry, W., and Melvill-Jones, G.: Influence of eye lid movement upon electrooculographic recording of vertical eye movements, Aerospace Med. **36**:855, 1965.
11. Baloh, R. W., and others: Quantitative measurement of smooth pursuit eye movements, Ann. Otol. Rhinol. Laryngol. **85**:111, 1976.
12. Morissette, Y., Abel, S. M., and Barber, H. O.: Optokinetic nystagmus in otoneurological diagnosis, Can. J. Otolaryngol. **3**:348, 1974.
13. Baloh, R. W., Hourubia, V., and Sills, A.: Eye-tracking and optokinetic nystagmus; results of quantitative testing in patients with well-defined nervous system lesions, Ann. Otol. Rhinol. Laryngol. **86**:108, 1977.
14. Gay, A. J., Newman, N. M., Keltner, J. L., and Stroud, M. H.: Eye movement disorders, St. Louis, 1974, The C. V. Mosby Co., pp. 50-55.
15. Jung, R., and Kornhuber, H. H.: Results of electronystagmography in man: the value of optokinetic, vestibular, and spontaneous nystagmus for neurologic diagnosis and research. In Bender, M. B., editor: The oculomotor system, New York, 1964, Harper & Row, Publishers, p. 438.

Chapter 7

Positional test

PROCEDURE

The technician performs the positional test by monitoring eye movements with the patient's eyes both open and closed and his head in various positions. The positions most commonly used are erect, supine, right lateral (RL), left lateral (LL), and head hanging (HH), as shown in Fig. 7-1. (The erect position has already been tested during the gaze test.) Eye movements should be recorded for at least 20 sec with the patient's eyes open and for at least another 20 sec with eyes closed in each head position. The purpose of the positional test is to detect nystagmus. When the technician is searching for nystagmus, the patient must be alert, and his eyes should be in the primary position.

The assumption of the positional test is that head position alone, and not head movement, is the cause of any nystagmus that is seen. It is, of course, impossible to change the head position of a patient without moving him, but the movement can be performed rather slowly to minimize the effect of this variable. Another variable that should be recognized is neck rotation. It is usually easier to attain the RL and LL positions by turning the patient's head rather than his whole body. However, if nystagmus is seen in either position when this method is used, is is necessary to have the patient assume the same position again, this time by turning the whole body onto its side rather than just the head. If in the second instance nystagmus disappears, it was caused at least partly by neck rotation and not by head position alone; however, this circumstance is quite uncommon.

NORMAL VARIATIONS

No normal individual has positional nystagmus with eyes open, but many have it with eyes closed in the presence of effective mental alerting. Barber and Wright[1] found that 92 of 112 normal individuals had nystagmus in at least one of a series of eight head positions with eyes closed. These authors judged nystagmus to be present whenever they recognized at least three consecutive beats in the ENG tracing. The nystagmus was always horizontal. In some people, it was *direction fixed;* that is, it beat in the same direction whenever it appeared. In others, nystagmus was *direction changing;* that is, it beat to the right in some head positions and to the left in others, although it never changed direction in a given head position.

142

Erect	Supine	Left lateral	Right lateral	Head hanging

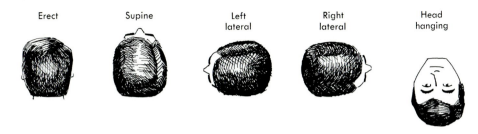

Fig. 7-1. Head positions used in the positional test. In subsequent illustrations of positional nystagmus, the specific positions are designated as follows: *S* (supine), *E* (erect), *RL* (right lateral), *LL* (left lateral), *HH* (head hanging), and not shown, *HHR* (head hanging right) and *HHL* (head hanging left).

Sometimes the nystagmus was *intermittent,* but sometimes it was *persistent;* that is, it was present as long as the head remained in a given position.

Since many normal people have horizontal positional nystagmus with eyes closed, one needs definite criteria for determining whether the finding is pathologic in a given patient. Based on their sample of 112 normal subjects, Barber and Wright established that horizontal positional nystagmus (eyes closed, effective mental alerting) is pathologic if:

1. Direction changes in single head position
2. It is persistent in three or more of five head positions
3. It is intermittent in four or more head positions
4. Slow-phase eye speed of the three strongest consecutive beats exceeds 6°/sec in any head position

Ninety-five percent of the normal individuals in this study failed to meet any of these criteria, so they represent the 95 percent limit of normal variability.

The tracing in Fig. 7-2 shows positional nystagmus with eyes closed in a normal individual. His nystagmus does not change direction in any head position, so he fails to meet criterion 1. He has persistent nystagmus in only one head position (RL), so he fails criterion 2. He has intermittent nystagmus in only two head positions (supine and LL), so he fails criterion 3.

To test criterion 4, it is necessary to calculate slow-phase eye speed for the three strongest nystagmus beats, which in this case appear in the RL position. An enlargement of the tracing containing those beats is hown in Fig. 7-3.

The slow-phase eye velocity during the strongest beat was 4°/sec (and the velocities of the two adjacent beats were approximately the same), so he fails to meet criterion 4. Therefore, his positional nystagmus with eyes closed in within normal limits.

ABNORMALITIES

Positional nystagmus with eyes open. Positional nystagmus with eyes open is always abnormal, hence of great importance. Sometimes tests for positional nys-

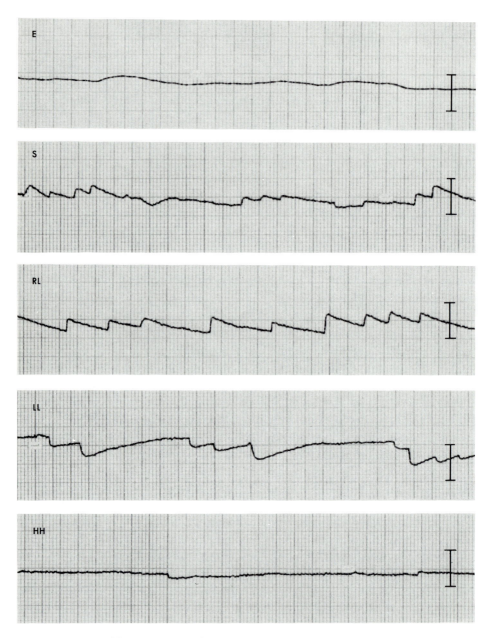

Fig. 7-2. Positional nystagmus of a normal individual. Eyes closed, bitemporal leads.

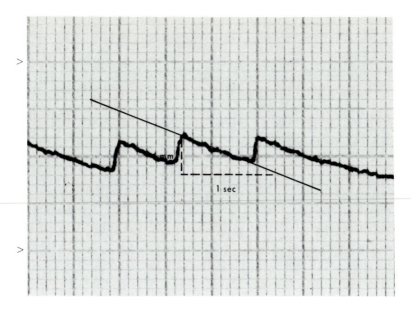

Fig. 7-3. Slow-phase eye speed is 4°/sec. For method of calculation see p. 161.

tagmus may be carried out in the presence of an active or convalescent illness of peripheral origin (for example, vestibular neuronitis), and direction-fixed nystagmus then may result. (The positional nystagmus in this case would be enhanced by eye closure.) More often the positional nystagmus is usually accompanied by little or no vertigo, and it persists for as long as the head position is maintained. Whether direction-fixed or direction-changing, it is then good evidence of CNS disease located within the posterior fossa (but see "Positional alcohol nystagmus," pp. 149-151).

Unlike paroxysmal positioning nystagmus (see pp. 153-158), whose production depends, in part at least, upon rapid movement from one head position to another, positional nystagmus of this type depends for its occurrence chiefly upon orientation of the head with respect to gravity. Thus it may be identified in the head-hanging-right (HHR) and head-hanging-left (HHL) positions in the Hallpike maneuver (see pp. 153-157).

Biemond and De Jong[2] have shown that positional nystagmus may result from cervical sensory root lesions in animals and man, but our clinical study leads us to the conclusion that positional nystagmus caused solely or predominantly by neck rotation is uncommon.

Direction-fixed positional nystagmus with eyes closed. The term *spontaneous nystagmus* has different meanings to different people. We prefer to define spontaneous nystagmus as nystagmus that is direction-fixed and beating with about the same intensity in all head positions when the eyes are closed. *Direction-fixed posi-*

tional nystagmus is differentiated from spontaneous nystagmus by a certain variability of intensity in different head positions, or by absence of the nystagmus in one or two positions, whether the eyes are open or closed. Testing a series of 112 normal individuals, Barber and Wright[1] found no spontaneous nystagmus within this restricted definition. Hence, they considered spontaneous nystagmus "always pathologic." On the other hand, as we have seen, direction-fixed positional nystagmus (with the eyes closed) is rather common in a normal population. Both spontaneous and direction-fixed positional nystagmus may result from peripheral lesions; as well, a given lesion may cause spontaneous nystagmus at one time and direction-fixed positional nystagmus at another (Fig. 7-5).

Fig. 7-4 shows portions of the record of a patient with right-sided Meniere's disease. When the patient is in the RL position, there is right-beating nystagmus with a slow-phase eye speed of about 7°/sec; when he is in the LL position, the nystagmus continues to beat to the right but with an intensity averaging about 20°/sec. When the patient is in the erect position, the intensity of the right-beating nystagmus is again about 5°/sec. This is direction-fixed positional nystagmus, and in patients with Meniere's disease, it may beat at different times either toward or

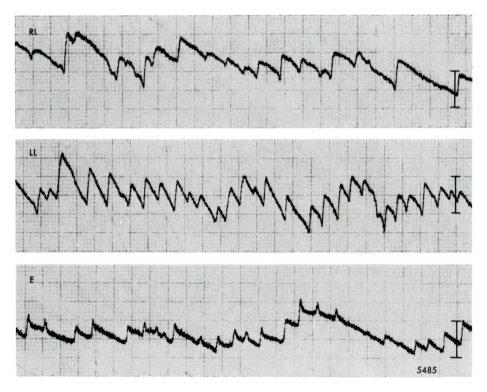

Fig. 7-4. Direction-fixed positional nystagmus of peripheral vestibular origin. Eyes closed, bitemporal leads.

away from the abnormal side. In other patients with acute peripheral vestibular diseases, especially vestibular neuronitis, direction-fixed positional nystagmus usually beats toward the normal side, but there are many exceptions, so one cannot infer laterality of lesion from the direction of direction-fixed positional nystagmus.

We have no example of direction-fixed positional nystagmus occurring from a case of well-defined CNS disease.

Spontaneous nystagmus may change with time into direction-fixed positional nystagmus. Fig. 7-5 is taken from a patient with typical right-sided vestibular neuronitis. There is strong left-beating nystagmus of identical speed in both the RL and LL positions (Fig. 7-5, *A*). Position has not altered the character or intensity of the nystagmus; hence, the definition of it is "spontaneous." One month later, coincident with considerable improvement and partial return of caloric function, left-beating nystagmus is still clear in the RL position but is of marginal intensity in the LL position (Fig. 7-5, *B*).

Direction-changing positional nystagmus with eyes closed. An example of direction-changing positional nystagmus with eyes closed is shown in Fig. 7-6. It was taken from the record of a patient with sudden cochlear loss caused by confirmed left-sided round window membrane rupture. There is marked right-beating nystagmus in the RL position and left-beating nystagmus in the LL position. Thus peripheral lesions may sometimes cause direction-changing positional nystagmus.

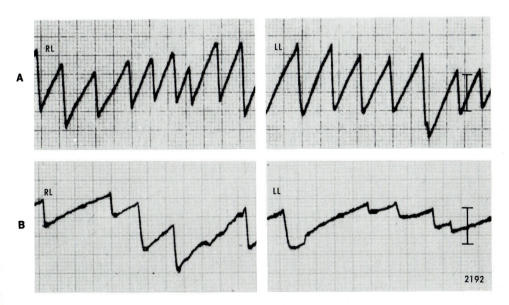

Fig. 7-5. Evolution of spontaneous nystagmus into direction-fixed positional nystagmus. **A,** Spontaneous nystagmus immediately after onset of right-sided vestibular neuronitis. **B,** Direction-fixed positional nystagmus 1 month later. **A** and **B,** Eyes closed, bitemporal leads.

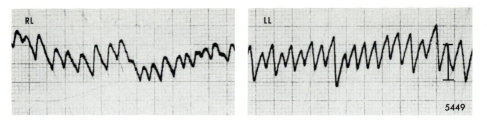

Fig. 7-6. Direction-changing positional nystagmus of peripheral vestibular origin. Eyes closed, bitemporal leads.

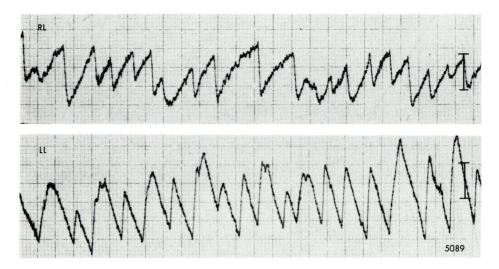

Fig. 7-7. Direction-changing positional nystagmus of CNS origin. Eyes closed, bitemporal leads.

The patient whose positional nystagmus is shown in Fig. 7-7 had brainstem infarction of clinically minor degree. On clinical examination with the patient's eyes open, no positional nystagmus was found. The tracing of this patient with eyes closed shows marked left-beating nystagmus in the RL position and intense right-beating nystagmus in the LL position. CNS disease may also produce direction-changing positional nystagmus and with considerably greater frequency than peripheral vestibular disorders. This case illustrates the rationale for conducting tests for positional nystagmus with the patient's eyes closed as well as open. Clearly pathologic positional nystagmus was identified in this instance only when the patient's eyes were closed.

Note that in Fig. 7-6, the tracings of a patient with a peripheral lesion, the nystagmus beats toward the floor in both lateral positions (geotropic), while in Fig. 7-7, the tracings of a patient with a central lesion, the nystagmus beats toward the

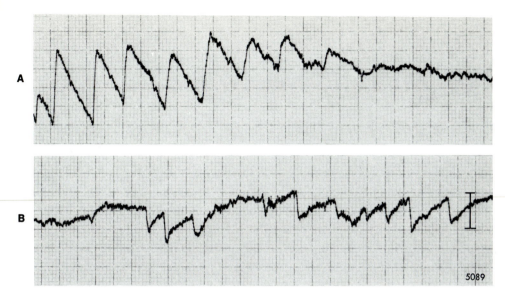

Fig. 7-8. Direction-changing positional nystagmus in a single head position *(HH)*. **B** is continuous with **A**. Eyes closed, bitemporal leads.

ceiling in both lateral positions (ageotropic). Provided that alcohol has not been ingested within the preceding 6 to 24 hours,[3] ageotropic direction-changing positional nystagmus suggests CNS localization.[4,5] In general, the ageotropic pattern is less common than the geotropic.

Direction-changing nystagmus in a single head position. The tracing in Fig. 7-8 is taken from the same patient as in Fig. 7-7. When the HH position is reached, right-beating nystagmus appears promptly, then soon declines to nil (Fig. 7-8, *A*). After a few seconds, nystagmus reappears (Fig. 7-8, *B*), this time beating to the left. This uncommon occurrence — positional nystagmus that changes in direction in a single head position — may be confined to CNS lesions.

Positional alcohol nystagmus (PAN). Direction-changing positional nystagmus is produced by large doses (at least 1 gm/kg of body weight) of ethyl alcohol (Fig. 7-9). The nystagmus appears approximately ½ hour after the alcohol ingestion and is most prominent when the head is in the RL and LL positions (Fig. 7-9, *A*). This nystagmus is called PAN I, and it is geotropic. It may be present with eyes open but is much stronger when they are closed. After 3 or 4 hours, PAN I disappears. Later, at least 5 hours after ingestion of alcohol, nystagmus appears again (Fig. 7-9, *B*). This nystagmus is called PAN II; it is ageotropic and persists for as long as 24 hours.

Many theories have been offered regarding the mechanism of PAN, but Money and others[3] have offered persuasive experimental evidence that suggests PAN is caused by a physical action of alcohol on the cupulae of the semicircular canals (Fig. 7-10). As alcohol diffuses into the labyrinth, it enters the cupulae be-

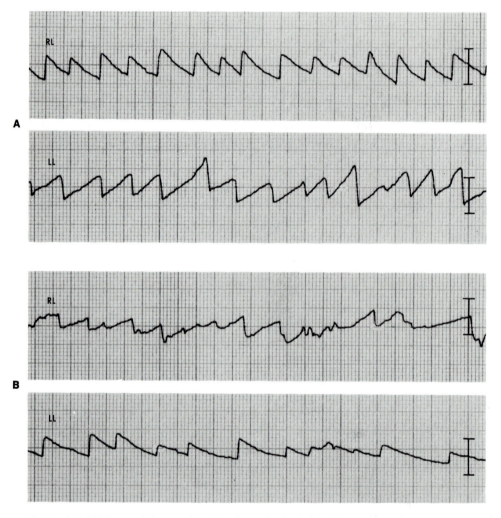

Fig. 7-9. A, PAN I, recorded approximately 1 hour after ingestion of 90 gm of alcohol. **B,** PAN II, recorded approximately 6 hours after alcohol ingestion. **A** and **B,** Eyes closed, bitemporal leads.

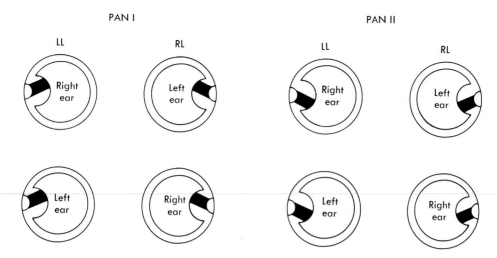

Fig. 7-10. Money and others' theory of the mechanism of PAN. For simplicity, only the lateral semicircular canals (viewed from above) are shown, and cupula deflections are grossly exaggerated.

fore it enters the endolymph. Since alcohol is less dense than endolymph, the cupulae also become less dense than endolymph and thus tend to rise, provoking PAN I when the head is in the RL and LL positions. When the alcohol diffuses into the endolymph, it makes the densities of cupulae and endolymph once again equal, and the nystagmus disappears. Finally, as alcohol diffuses out of the labyrinth, it leaves the cupulae before leaving the endolymph, making the cupulae temporarily heavier than the endolymph and provoking PAN II.

PITFALLS

Since direction-changing positional nystagmus can be provoked by alcohol ingestion, one must not attribute this abnormality to an organic lesion until alcohol has been ruled out as a cause. One rarely observes PAN I in the clinical ENG examination, because few patients would come for testing while intoxicated. However, one is quite likely to see PAN II.

Several other variables can influence the results of the positional test. One variable is *alertness*. As noted previously, it is important for the technician to keep the patient alert by some task, such as mental arithmetic, when looking for nystagmus when the patient's eyes are closed, because lack of alertness causes nystagmus suppression. The problem is increased when the patient is taking CNS-depressant medication (p. 111).

Another variable is *direction of gaze*. When placed in the RL or LL positions with eyes open, patients tend to look either at the ceiling or at the floor. If the patient has gaze nystagmus, it would then appear in the tracing and might be erroneously interpreted as positional nystagmus. The technician should check to see that the patient is looking straight ahead during the positional test.

Table 7-1. Positional nystagmus

Finding	Localization	Comment
Nystagmus with eyes open (pp. 143-145)		
Direction-fixed	Usually CNS	
Direction-changing	Always CNS	Rule out PAN II
Nystagmus with eyes closed (pp. 145-149)		
Significant of abnormality (pp. 142-143)		
Direction-fixed	"Always" peripheral	
Direction-changing	Usually CNS	If nystagmus ageotropic, likelihood of CNS lesion increased, rule out PAN II
Direction-changing nystagmus in a single head position	CNS	

The RL or LL positions may be attained with the body supine by turning only the head to either side or by turning the patient on to each side without neck rotation. Our clinical study indicates that one maneuver is the practical equivalent of the other; neck rotation is rarely the dominant factor in production of positional nystagmus.

REFERENCES

1. Barber, H. O., and Wright, G.: Positional nystagmus in normals, Adv. Otorhinolaryngol. **19**:276, 1973.
2. Biemond, A., and De Jong, J.M.B.V.: On cervical nystagmus and related disorders, Brain **92**:437, 1969.
3. Money, K. E., Myles, W. S., and Hoffert, B. M.: The mechanism of positional alcohol nystagmus, Can. J. Otolaryngol. **3**:302, 1974.
4. Kornhuber, H. H.: The vestibular system; functional and clinical correlates, lecture notes for course on oculomotor-vestibular disorders, University of Toronto, 1973.
5. Barber, H. O.: Nystagmus in tumours of the eighth nerve. In Naunton, R. F., editor: The vestibular system, New York, 1975, Academic Press, Inc., p. 433.

Chapter 8

Paroxysmal positioning nystagmus

HALLPIKE MANEUVER

Severe positional vertigo is a common clinical complaint. The most frequent form of nystagmus, and often the sole physical sign that accompanies this complaint, is positioning nystagmus. We specify position*ing* rather than position*al* nystagmus because rather rapid movements from one position to another may be needed to elicit it. A special procedure, the Hallpike maneuver,[1-3] is required for identifying positioning nystagmus; this procedure should be performed on all patients.

Procedure

We advocate that the technician perform the Hallpike maneuver while recording eye movements. The physician also performs the maneuver, preferably while reducing ocular fixation with Frenzel's glasses in a darkened room. Paroxysmal positioning nystagmus is an important physical sign, yet one that varies from time to time, and the practice of both physician and technician seeking to identify the sign seems justified for these reasons.

A suitable procedure for the technician is to conduct the tests with the patient's eyes open in the primary position, because a common form of positioning nystagmus is largely rotatory, which is not always well recorded by the ENG method and is better detected by visual observation. However, horizontal and vertical forms of positioning nystagmus do occur,[4,5] and even combined forms are not rare. Therefore, ENG recording during the test is worthwhile.

Fig. 8-1 illustrates a suitable test procedure. The patient is seated in good light near the end of an examining table, with sufficient space at its head for maneuvering. The examiner looks for nystagmus with the patient sitting, then moves the head, neck, and body briskly into the HHL position. The shoulders are flat at the end of the table, and the neck is rotated to one side; *the degree of rotation should vary with the flexibility of the neck and should not be excessive, especially in elderly patients*. The examiner supports the undersurface of the patient's head with one hand, while using the other hand to provide an ocular fixation point. This position is held for 10 to 15 sec. The patient is then briskly returned to the erect posi-

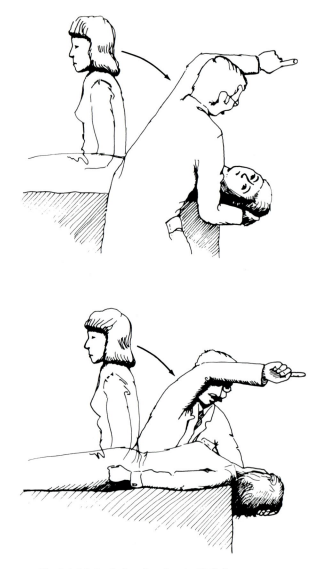

Fig. 8-1. Method of performing the Hallpike maneuver.

tion, and the examiner again looks for nystagmus for 10 to 15 sec. The next sequence is the HHR position, again followed by the erect position.

Paroxysmal positioning nystagmus is subject to suppression by ocular fixation.
Thus, an alternative laboratory technique is to perform the Hallpike maneuver
and record eye movements with eyes closed. Probably paroxysmal positioning
nystagmus is recorded at times by this means when it would be absent with eyes
open, but it is more difficult to obtain a stable ENG baseline, the opportunity for
visual observation of the patient's eyes is lost, and persistent, nonparoxysmal

forms of positional nystagmus (p. 145) occurring with the eyes open may not be detected. The experienced technician may choose to adopt this variation of technique if he is confident that the physician will also perform the maneuver as part of the otoneurologic physical examination.

Normal variations

In normal individuals, the Hallpike maneuver provokes no vertigo and no nystagmus with the eyes open, although a few beats are sometimes seen in the ENG tracing when one is recording with the patient's eyes closed.

Abnormalities

The most common form of nystagmus provoked by the Hallpike maneuver is *positioning nystagmus of the benign paroxysmal type.*[2,6] This nystagmus appears after the patient has been placed in either the HHR or HHL position, sometimes both. It is horizontal-rotatory and beats toward the undermost ear. Often it briefly reappears, beating in the opposite direction when the patient resumes the sitting position. The salient features of positioning nystagmus of the benign paroxysmal type are contained in Table 8-2, p. 158.

Identification of this rather specific type of positioning nystagmus is important because it is the most common form in clinical practice, it usually gives laterality (undermost ear),[1,2,8] and it nearly always[5,7] denotes harmless inner ear disease.

Fig. 8-2 shows the record of a patient with classic (idiopathic) positioning nystagmus of the benign paroxysmal type when tested with the Hallpike maneuver. With the patient in the HHR position, a predominantly rotatory nystagmus beating counterclockwise after a few seconds' latency is easily visible. The nystagmus quickly reached a crescendo and then declined to nil in about 10 sec. Note that while the patient had predominantly rotatory nystagmus beating counterclockwise, the bitemporal and vertical leads show apparent nystagmus beating obliquely, a little upward but mainly to the left. Positioning nystagmus of this type can usually be recorded by ENG, with the patient's eyes either open or closed, and there is almost always an upbeating component in the tracing from the vertical leads.

Note the direction of beat in the bitemporal leads. When rotatory positioning nystagmus of the benign paroxysmal type is recorded by ENG, the horizontal nystagmus appears to beat in an "unexpected" direction. That is, in the HHR position, the bitemporal leads show nystagmus beating to the *left,* and in the HHL position, to the *right.* This is true with eyes both open and closed.

Paroxysmal positioning nystagmus may occur in both HHL and HHR positions, rather than in just a single lateral position. The characteristic rotational nystagmic eye movements are clockwise in HHL and counterclockwise in HHR. Fig. 8-3 is an example, in a patient with right temporal bone fracture. The Hallpike maneuver produced characteristic paroxysmal nystagmus in each head position, with latency of 1 to 2 sec, rapid attainment of peak intensity, and slower de-

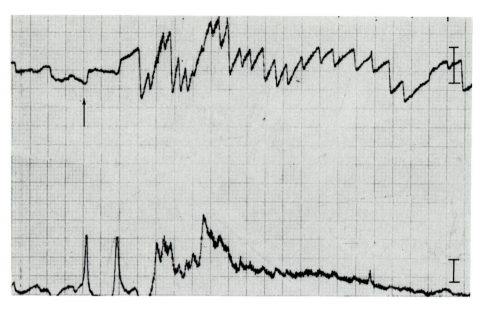

Fig. 8-2. Positioning nystagmus of benign paroxysmal type. Eyes open; upper tracing, bitemporal leads; lower tracing, vertical leads. The *HHR* position is attained at the arrow.

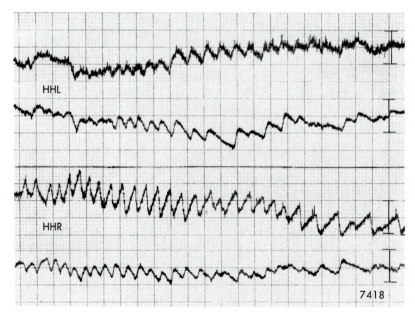

Fig. 8-3. Bilateral paroxysmal positioning nystagmus in patient with right-sided temporal bone fracture. *HHL* and *HHR,* Top tracing, bitemporal leads; lower tracing, vertical lead, right eye. See text. (From Longridge, N. S., and Barber, H. O.: Bilateral paroxysmal positioning nystagmus, J. Otolaryngol. 7: 395, 1978.)

cline of nystagmus to nil in about 13 to 15 sec. The nystagmus here is more intense in HHR position than in HHL position, and it was accompanied by vertigo in each case.

Longridge and Barber[8] found bilateral paroxysmal positioning nystagmus in 15 percent of 114 patients with classic benign positional vertigo. Head injury and/or CNS abnormality were prominent as antecedent causes.

The Hallpike maneuver is also effective in identification of positional nystagmus (pp. 143-145). This type of nystagmus differs markedly from nystagmus of the benign paroxysmal type; it appears without latency, is horizontal, oblique, or vertical (not rotatory), lasts for a long time (usually as long as the head position is held), is accompanied by little or no vertigo, and fatigues little or not at all upon repeated testing. This type of positional nystagmus is of particular importance because it usually denotes significant CNS disease.

Table 8-1. Summary of abnormalities observed in the positional test and Hallpike maneuver

Abnormality	Significance	Comment
Positional test		
Nystagmus with eyes open (pp. 143-145)		
Direction-fixed	Usually CNS lesion	
Direction-changing	Always CNS lesion	Rule out PAN II
Nystagmus with eyes closed* (pp. 145-149)		
Direction-fixed	Usually peripheral vestibular lesion	
Direction-changing	Usually CNS lesion	If nystagmus is ageotropic, likelihood of CNS lesion increased; rule out PAN II
Direction-changing nystagmus in a single head position	CNS lesion	
Hallpike maneuver		
Benign paroxysmal-type positioning nystagmus (pp. 155-158)		
Unilateral	Usually peripheral vestibular lesion (undermost ear)	Response must be delayed in onset, transient, fatigable, and accompanied by vertigo
Bilateral	Peripheral vestibular lesion (both ears) or CNS lesion	
Any other nystagmus (pp. 143-144)	Usually CNS lesion	Immediate onset, persists, nonfatigable, little or no vertigo

*To be abnormal, positional nystagmus with eyes closed must meet one or more of the criteria listed on p. 143.

Table 8-2. Paroxysmal positioning nystagmus (Hallpike maneuver)

Finding	Significance	Comment
Unilateral	Usually peripheral, usually undermost ear	Response must be delayed in onset, transient, fatigable, and accompanied by vertigo
Bilateral	Peripheral vestibular (both ears) or CNS	

Pitfalls

Positioning nystagmus of the benign paroxysmal type is fatigable; therefore, the Hallpike maneuver must be done correctly on the first attempt. The response may be absent the second time. If the examiner intends to perform the Hallpike maneuver, he should do it before other manipulations of the patient that could trigger the response. The examiner should prepare the patient beforehand, telling him what he will do and what he wants to see. The patient who has benign paroxysmal vertigo is usually reluctant to perform the Hallpike maneuver. Even when convinced of its necessity, he is apt to become agitated and tightly close his eyes once the vertigo starts, making visual observation or recording of nystagmus impossible. He must be told to keep his eyes open at all costs, if this is the test technique in use.

SUMMARY

The summary of the significance of paroxysmal positioning nystagmus is given in Table 8-2.

REFERENCES

1. Dix, M. R., and Hallpike, C. S.: The pathology, symptomatology, and diagnosis of certain common disorders of the vestibular system, Proc. R. Soc. Med. **45:**341, 1952.
2. Cawthorne, T.: Positional nystagmus, Ann. Otol. Rhinol. Laryngol. **63:**481, 1954.
3. Barber, H. O.: Positional vertigo and nystagmus, Otolaryngol. Clin. North Am. **6:**169, 1973.
4. Stahle, J., and Terins, J.: Paroxysmal positional nystagmus, Ann. Otol. Rhinol. Laryngol. **74:**69, 1965.
5. Harrison, M. S., and Ozsahinoglu, C.: Positional vertigo, Arch. Otolaryngol. **101:**675, 1975.
6. Barber, H. O.: Positional nystagmus, especially after head injury, Laryngoscope **74:**891, 1964.
7. Riesco-MacClure, J. S.: ¿Es el vértigo aural de origen exclusivamente periférico? Rev. Otorhinolaringol. **17:**42, 1957.
8. Longridge, N. S., and Barber, H. O.: Bilateral paroxysmal positioning nystagmus, J. Otolaryngol. **7:**395, 1978.

Chapter 9

Caloric test

Of all the ENG tests, the caloric test is the most difficult and time-consuming for both the technician and the patient; yet it is of great value to the otoneurologist.

PROCEDURE
Irrigations

The caloric test is performed by irrigating each external auditory canal twice — once with warm water (or air) and once with cool water (or air) — and recording each of the four provoked nystagmus responses. The strength of the caloric stimulus depends on a number of variables. The examiner can control three of them: volume, duration, and temperature of the irrigating medium. Commonly used specifications for both water and air caloric stimuli are listed in Table 9-1. These values produce stimuli that provoke clear nystagmus responses in most people when their eyes are closed, but they do not provoke responses strong enough to cause severe symptoms of motion sickness. When the specifications listed in Table 9-1 are used, water and air caloric stimuli produce responses of approximately equal intensity.[1]

The examiner cannot control the other variables that influence the strength of the caloric stimulus. The most important of these is the anatomy of the ears. For example, the patient who has wide and straight external auditory canals will, in general, receive a stronger stimulus and therefore have a stronger response than one who has narrow, tortuous canals. Because of anatomic variability and other uncontrollable factors, the strength of the caloric stimulus that reaches the labyrinth varies from individual to individual, even though the strength of the stimulus that enters the external auditory canal is the same for everyone. For this reason, the range of response strength is extremely wide, and only gross comparisons can be made among patients. However, it is assumed that the two ears of a given patient do not differentially affect stimulus strength, and thus that the strengths of the stimuli that reach the two labyrinths of a given individual are equal. This is the fundamental assumption underlying the caloric test.

The caloric test is performed with the patient in the supine position with the head flexed 30°, so that the plane of the lateral semicircular canals is vertical. When the head is in this position, each caloric irrigation stimulates the lateral

159

Table 9-1. Standard specifications for water and air caloric stimuli

Variable	Irrigating medium	
	Water	Air
Volume	250 ml	8 liters
Duration of flow	30 sec	60 sec
Warm temperature	44° C	50° C
Cool temperature	30° C	24° C

semicircular canal of the irrigated ear. The mechanism of caloric stimulation is described on pp. 28-29.

Before beginning the test, one should tell the patient what to expect. Both the water and air irrigations sound quite noisy. When the warm temperature is used, it feels hot, but it is not painful. Some patients have little or no vertigo from the stimulus, some have a vague feeling of light-headedness, but most have a clear illusion of themselves or their surroundings (or both) turning. Patients also report a variety of other motions, probably because vestibular receptors other than the lateral canal are also being stimulated by the temperature wave. Some patients are alarmed when they experience vertigo, but they are nearly always reassured if they are told that vertigo is a normal response to the irrigation, that it will probably be mild, and that it will last for only a minute or two.

When the patient is ready for the first irrigation, the technician recalibrates the recording system, then asks the patient to close his eyes and keep them closed. The recorder should be allowed to run for 15 sec before the irrigation begins (the patient should be given alerting tasks in this period) and during the irrigation itself, because this is an excellent time for observing any preexisting nystagmus. Nystagmus that may have been suppressed during the gaze and positional tests because of the patient's lack of alertness will probably be released from suppression at this time, because the patient is alert as he anticipates and then experiences the irrigation. After looking for preexisting nystagmus, the technician places the irrigating tip into the external auditory canal. Direct visual observation of the tip is desirable, because the patient often jerks as he feels the first blast of water in his ear and sometimes dislodges the tip, which the technician must then reposition. (An otologist's headlight serves well for this purpose.) The technician then begins the irrigation. Nystagmus response (and vertigo) usually appears near the end of the irrigation, then builds an intensity for approximately 30 sec, and declines thereafter. It is vitally important that the patient be kept alert during the peak nystagmus response, because this portion of the tracing will be used later to obtain a quantitative estimate of response strength. After the response has begun to decline (approximately 90 to 100 sec after the beginning of the irrigation), the patient is asked to open his eyes and fixate on the technician's finger or a spot on the ceiling. After 10 sec, the patient is asked to close his eyes again, and the recording is continued until nystagmus is no longer identifiable or the record has returned to its pretest state. It is then necessary to wait several minutes for the labyrinth to reach body temperature again before beginning the next irrigation. A wait of 5 min from the end of one irrigation to the beginning of the next is adequate. The recording system should be recalibrated before the next irrigation.

Calculations

Parameters of response. After all four irrigations have been completed, the technician needs to obtain a quantitative estimate of the strength of each of the four responses. In the past, three indexes of response strength have been used by otoneurologists. The first is *duration of the nystagmus response,* which is usually defined as the interval between the beginning of the irrigation and the last beat of nystagmus. It is often difficult to identify the last beat of nystagmus; therefore, arbitrary criteria for the end of the response, such as the last time two nystagmus beats are seen within a 5-sec interval, have been suggested. The second index is *peak nystagmus frequency,* which is usually defined as the average frequency of nystagmus beats during the 10-sec interval in which nystagmus is most intense. The third index of response strength, which is most widely used, is the one we advocate to the exclusion of the others. It is *maximum slow-phase eye speed,* which is defined as the average slow-phase eye speed during the 10-sec interval in which the response is most intense (usually between 55 and 65 sec after the beginning of the irrigation, although it must be determined by inspection of the tracing in each case). Perhaps the simplest way to obtain an estimate of maximum slow-phase eye speed is to measure eye speed of the three strongest beats of nystagmus within the chosen 10-sec interval and then to average these values. The procedure for calculating slow-phase eye speed is described on pp. 78-81.

Unilateral weakness. After the technician has calculated maximum slow-phase eye speed for each of the four caloric responses, he can calculate unilateral weakness, or the amount by which the two responses provoked by right ear irrigations differ in intensity from those provoked by left ear irrigations. Unilateral weakness is calculated by the following formula:

$$\text{Unilateral weakness} = \frac{(RW + RC) - (LW + LC)}{(RW + RC + LW + LC)} \times 100$$

where RW is peak slow-phase eye speed of the response following the right ear-warm temperature irrigation, RC is peak response for the right ear-cool temperature irrigation, LW is the peak response for the left ear-warm temperature irrigation, and LC is the peak response for the left ear-cool temperature irrigation. Unilateral weakness is thus the amount by which the responses to irrigation of the two ears differ, expressed as a percentage of the sum of all four responses.

Directional preponderance. The same four elements are used to calculate directional preponderance, which represents the difference in intensity between the two right-beating nystagmus responses (provoked by right ear-warm temperature and left ear-cool temperature irrigations) and the two left-beating responses (provoked by left ear-warm temperature and right ear-cool temperature irrigations). Directional preponderance is calculated by the following formula:

$$\text{Directional preponderance} = \frac{(RW + LC) - (LW + RC)}{(RW + LC + LW + RC)} \times 100$$

where the abbreviations are the same as those used in the formula for calculating unilateral weakness.

In this book we designate a unilateral weakness according to the side of the *weaker* response and a directional preponderance according to the direction of the *stronger* responses. Thus, the abbreviation *UWR7%* means that the right side's caloric response is 7 percent weaker than that of the left; *DPL20%* means that there is a 20 percent preponderance of left-beating over right-beating caloric nystagmus.

Fixation index. The fixation index (FI) is a measure of the effectiveness of visual fixation in suppressing caloric nystagmus. It is calculated by the following formula:

$$FI = \frac{SPES\ (EO)}{SPES\ (EC)}$$

where *SPES (EO)* is the slow-phase eye speed of two or three representative beats occurring while the eyes are open and fixating, and *SPES (EC)* is the slow-phase eye speed of two or three representative beats occurring just before the eyes are opened. The FI should be calculated for at least one right-beating and one left-beating caloric response.

NORMAL VARIATIONS

Table 9-2 gives data on normal water and air caloric values from different clinical laboratories. The figures vary widely. This results from a combination of variation in instrumentation, test technique, procedure, and analysis of records. No one investigator duplicates precisely the study of another. The reader should realize that normal values, while essential for daily clinical work, are only a guide.

Normal individuals and patients with peripheral vestibular disease exhibit marked suppression and sometimes abolition of caloric nystagmus when they open their eyes and fixate. Alpert[10] (calculating SPES[EC] in a slightly different way from that discussed above) has reported that an FI of 0.6 represents the 95 percent limit of normal variation. His estimate is based on a sample of thirty individuals. The same value for FI at Sunnybrook Medical Centre is 0.7.

Fig. 9-1 shows the caloric responses of a normal person. The first four tracings are segments containing the peak nystagmus responses to the four irrigations, and the fifth is a segment containing the effect of visual fixation (*arrow*) on the intensity of the response. It is apparent that the four responses beat in the appropriate directions and that their intensities are approximately the same. The latter impression is confirmed when unilateral weakness and directional preponderance are

Table 9-2. Normal caloric values*

Source	N	Water						Air					
		Warm		Cool		R/L	DP	Warm		Cool		R/L	DP
		Mean	Range	Mean	Range	percent†	percent†	Mean	Range	Mean	Range	percent†	percent‡
Barber and Wright[3]	114	35	11-80†	28	6-50†	25	23	37	11-85	30	10-46	25	23
Barber and Wright[3]	24							20	15-25†	15	10-20†		
Ford and Stockwell[2]	8	22	16-28†	17	11-23†								
Baloh and Honrubia[4]	44	21	6-68	15	5-40	22	28						
Hamersma[5] Jongkees and Philipszoon[6]	47	23	8-52	22	9-46	15	18						
Mehra[7]	31	26	10-52†	21	3-39†								
Henriksson[8]	25	29	8-65	29	8-45								
Capps and associates[1]	10	16	6-26†	18	8-28†			15	3-27†	17	7-27†		
Benitez and associates[9]	30							22	6-38	21	5-37	16	13

*Values given for mean and range refer to raw slow component velocity at period of peak response in degrees per second.
†95 percent limits.
‡Estimate.

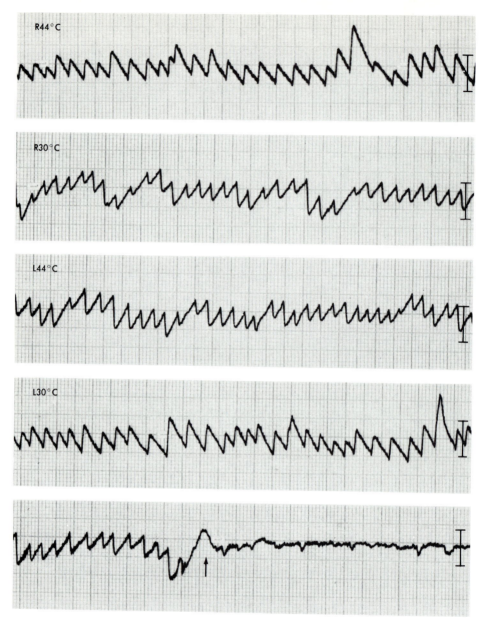

$$\text{Unilateral weakness} = \frac{(18 + 21) - (23 + 22)}{18 + 21 + 23 + 22} = \frac{-6}{84} = 7\% \text{ on the right}$$

$$\text{Directional preponderance} = \frac{(18 + 22) - (23 + 21)}{18 + 22 + 23 + 21} = \frac{-4}{84} = 5\% \text{ to the left}$$

$$\text{Fixation index} = \frac{1}{20} = 0.05$$

Fig. 9-1. Normal caloric responses. Bitemporal leads.

NOTE: In this book we designate a unilateral weakness according to the side of the *weaker* responses and a directional preponderance according to the direction of the *stronger* responses. However, the reader should be aware that the terminology for designating the side of a unilateral weakness is not standardized. Sometimes unilateral weakness is designated according to the side of the stronger responses.

calculated. Both are well within the normal range. The nystagmus of this individual is strongly suppressed when he opens his eyes and fixates, and indeed his FI is within the normal range of variation.

ABNORMALITIES
Unilateral weakness and directional preponderance

Fig. 9-2 shows the peak caloric responses of a patient who has a unilateral weakness on the left side. The weakness is apparent by inspection of the tracings and is confirmed when unilateral weakness is calculated. The directional preponderance of this patient is well within the normal range of variation.

Fig. 9-3 shows a directional preponderance to the right. Right-beating responses are clearly stronger than left-beating responses. This is confirmed when directional preponderance is calculated. Unilateral weakness is well within the normal range. This individual also had a preexisting nystagmus to the right when his eyes were closed. A person with preexisting nystagmus usually displays a directional preponderance in the same direction, but sometimes a directional preponderance is present without preexisting nystagmus.

The tracings shown in Fig. 9-4 illustrate a considerably more complicated but not uncommon picture. This patient has both unilateral weakness on the left side and a directional preponderance to the right. The response to warm irrigation of the right ear beats in the same direction as the preponderance; therefore, it is much stronger than the response to cool irrigation of the same ear, which opposes the preponderance. The response to warm irrigation of the left ear, which should beat to the left, is so weak that it cannot overcome the right preponderance; thus no nystagmus is seen. The response to cool irrigation of the left ear, which beats to the right, is also weak.

A significant unilateral weakness is a most important ENG finding for the otoneurologist. It commonly signifies a peripheral vestibular lesion on the weak side. No other ENG finding so positively identifies and lateralizes a peripheral vestibular lesion.

A significant directional preponderance, on the other hand, is of little clinical value. It is a sign that something is probably wrong, but it offers no localizing value, because it accompanies a wide variety of both peripheral vestibular and CNS disorders.[11]

Bilateral weakness

In some patients, the caloric responses of both ears are very weak or absent. When each of the four caloric irrigations has peak slow component velocity lower than the lower value of the range of caloric responses noted in Table 9-2, bilateral weakness (BW) is present. The particular values chosen to define this occurrence clearly depend on the range of normal values adopted. For example, in Barber and Wright's normal series (Table 9-2) for water calorics, the low normal value for warm is 11°/sec, for cool 6°/sec. Therefore, provided responses are bi-

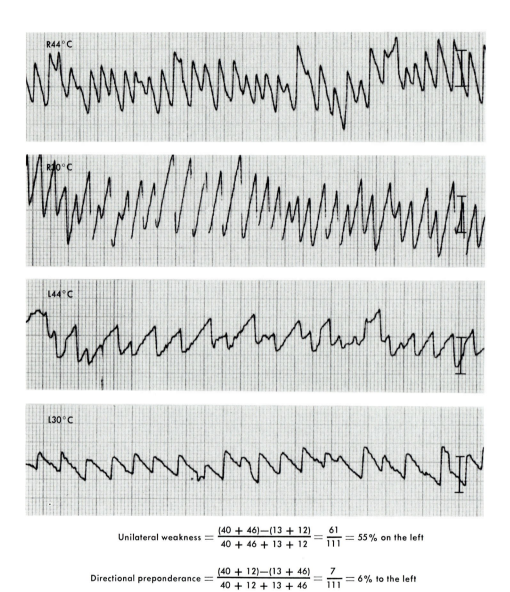

$$\text{Unilateral weakness} = \frac{(40 + 46)-(13 + 12)}{40 + 46 + 13 + 12} = \frac{61}{111} = 55\% \text{ on the left}$$

$$\text{Directional preponderance} = \frac{(40 + 12)-(13 + 46)}{40 + 12 + 13 + 46} = \frac{7}{111} = 6\% \text{ to the left}$$

Fig. 9-2. Peak caloric responses of a patient who has a unilateral weakness on the left. Bitemporal leads.

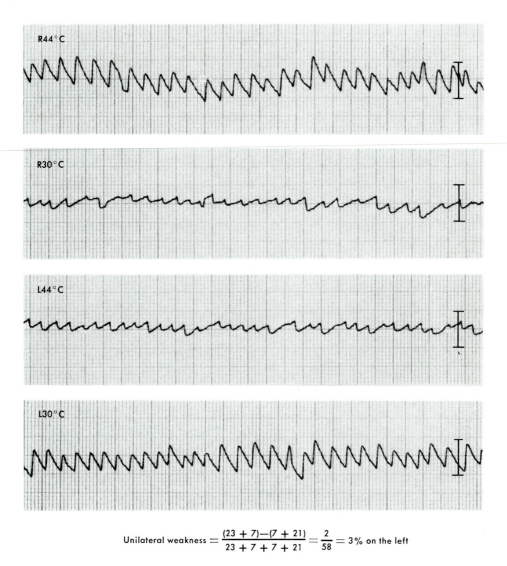

R44°C

R30°C

L44°C

L30°C

$$\text{Unilateral weakness} = \frac{(23+7)-(7+21)}{23+7+7+21} = \frac{2}{58} = 3\% \text{ on the left}$$

$$\text{Directional preponderance} = \frac{(23+21)-(7+7)}{23+21+7+7} = \frac{30}{58} = 52\% \text{ to the right}$$

Fig. 9-3. Peak caloric responses of a patient who has a directional preponderance to the right. Bitemporal leads.

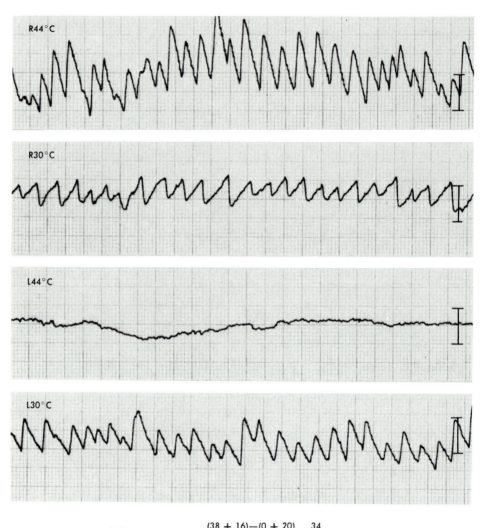

$$\text{Unilateral weakness} = \frac{(38 + 16)-(0 + 20)}{38 + 16 + 0 + 20} = \frac{34}{74} = 46\% \text{ on the left}$$

$$\text{Directional preponderance} = \frac{(38 + 20)-(0 + 16)}{38 + 20 + 0 + 16} = \frac{42}{74} = 57\% \text{ to the right}$$

Fig. 9-4. Peak caloric responses of a patient who has a unilateral weakness on the left and a directional preponderance to the right. Bitemporal leads.

laterally symmetrical, BW would be present when the sum of the four caloric irrigations (2×10 for warm $+ 2 \times 5$ for cool $= 30°/sec$) is about 30°/sec or less.

An example of bilateral weakness in a patient who had received streptomycin therapy is shown in Fig. 9-5. Both streptomycin and gentamicin are ototoxic in sufficient dosage, especially when serum levels are high because of impaired renal clearance. Bilateral caloric reduction occurs with other peripheral vestibular lesions as well as CNS lesions. Other peripheral causes that we have identified are bilateral temporal bone fracture, bilateral eighth nerve tumors, bilateral Meniere's disease, and Cogan's syndrome.[12,13] Simmons[14] gives a comprehensive list of causes that includes a number of primary CNS lesions.

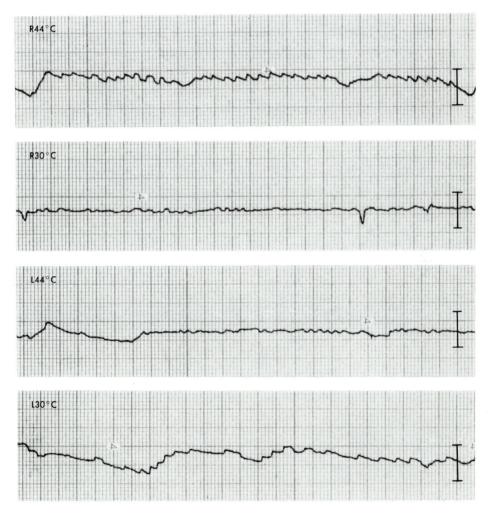

Fig. 9-5. Peak caloric responses of a patient who has a bilateral weakness. Bitemporal leads.

Hyperactive response

Caloric responses may exceed the upper limit of normal variation, 50°/sec slow-phase eye speed for cool irrigations and 80°/sec for warm irrigations.[3] Such findings are uncommon in our experience.

Hyperactive responses may occur if the caloric transfer qualities of the stimulated ear are greatly enhanced, such as when a mastoidectomy cavity is present or when the tympanic membrane is perforated, atrophic, or retracted. Otherwise, these responses are generally explained as resulting from reduction of normal CNS (mainly cerebellar) inhibition of caloric activity.[15,16] Fredrickson and Fernandez[17] produced hyperactive caloric responses (among other abnormalities) in cats with lesions of the cerebellar nodulus and adjacent vestibular structures in the floor of the fourth ventricle. Remember that, in earlier animal studies, caloric activity was recorded with eyes open. Thus, "hyperresponsiveness" in this test condition may have actually represented failure of fixation suppression due to the cerebellar lesion. Even so, increased caloric responses occur in patients who have no evidence of intracranial disease; nervousness or overalertness may well be responsible in such cases.

Failure of fixation suppression

In all normal individuals, all patients with peripheral vestibular disorders, and in some patients with CNS disorders, caloric nystagmus is suppressed by visual fixation. However, in some patients with CNS disorders, nystagmus intensity with eyes open nearly equals, matches, or exceeds that with eyes closed. This effect is known as failure of fixation suppression[18] and is good evidence of CNS localization. The vestibulocerebellum, especially the flocculus and its connections, has been shown to mediate this effect.[19] The finding may occur only in one direction of nystagmus.

The tracing in Fig. 9-6, *A*, taken from a patient with benign positional vertigo, shows normal fixation suppression. The slow-phase eye speed of nystagmus with eyes closed is about 48°/sec; with eyes open and fixed, it is about 14°/sec. The FI is about 0.3, which is within normal limits. The tracing in Fig. 9-6, *B*, taken from a patient with pontine infarction, shows failure of fixation suppression. Slow-phase eye speed with eyes closed is 24°/sec; with eyes open, 24°/sec. The FI is therefore 1.0, which is outside the normal range of variation.

Failure of fixation suppression may occasionally be so marked that caloric nystagmus appears only with eyes open and not at all with eyes closed. In diffuse metabolic brain disease or barbiturate or phenytoin (Dilantin) intoxication, caloric nystagmus may be entirely absent with eyes closed at time of expected peak activity, despite vigorous efforts at alerting. The nystagmus then appears only with ocular fixation, eyes open. Drug interference with reticular formation activity is probably responsible for this unusual occurrence.

Fig. 9-7 is the caloric record of a patient with Wernicke's encephalopathy. In Fig. 9-7, *A*, nystagmus is absent with eyes closed but appears promptly and in the expected direction when the eyes are open and fixed. Fig. 9-7, *B*, shows that, after

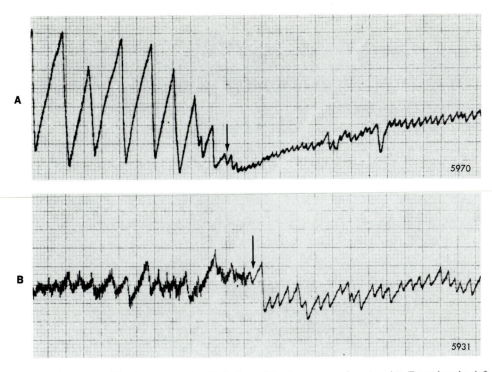

Fig. 9-6. **A,** Normal fixation suppression. **B,** Failure of fixation suppression. **A** and **B,** Eyes closed to left of arrow and open to right of arrow, bitemporal leads.

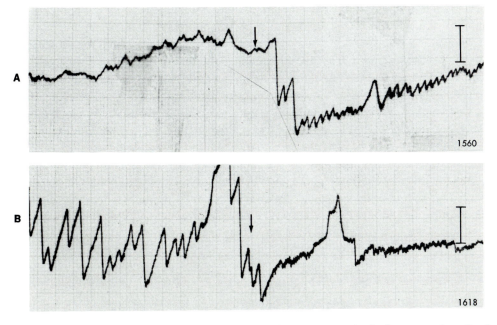

Fig. 9-7. **A,** Failure of fixation suppression. **B,** Normal fixation suppression in the same patient after 1 month of treatment with thiamine chloride. **A** and **B,** Eyes closed to left of arrow and open to right of arrow, bitemporal leads.

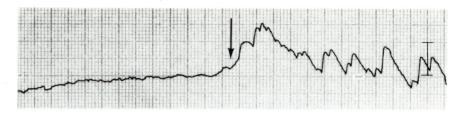

Fig. 9-8. Caloric nystagmus absent with eyes closed, mental alerting (left of arrow), present with eyes open and fixated (right of arrow). Barbiturate intoxication in epilepsy. Bitemporal leads.

1 month's treatment with thiamine chloride, caloric nystagmus to the same stimulus is normal with eyes closed and becomes markedly suppressed with fixation.

Fig. 9-8 is a caloric record (left ear, cool-air, time of expected peak response) of a patient with chronic barbiturate intoxication from treatment of epilepsy. Nystagmus is absent with eyes closed and appears promptly when the eyes are open and fixated.

Premature caloric reversal

The caloric response generally reaches peak intensity between 45 and 90 sec after the start of irrigation, and the nystagmus then slowly declines until it stops after about 200 sec. If the recorder is allowed to run, weak secondary nystagmus sometimes reappears, beating in the opposite direction, then it too declines to nil. This phenomenon is called caloric reversal.[20]

When the caloric reversal occurs too early and is particularly strong, it denotes a disorder of CNS (probably cerebellar system) mechanisms that modulate the caloric response, and it may be a feature of posterior fossa lesions. Fig. 9-9, an example of this occurrence, is taken from a patient with brainstem infarction who also had unilateral internuclear ophthalmoplegia. There was no preexisting nystagmus before caloric stimulation. Fig. 9-9, *A*, shows that response is in the expected direction for about 60 sec, but soon afterward it declines to nil. Fig. 9-9, *B*, shows that right-beating nystagmus commences after 69 sec and gains in intensity. Fig. 9-9, *C*, shows that it continues beating actively in the inverted direction.

It is important not to confuse premature caloric reversal with resumption of preexisting nystagmus, which, of course, might well beat in a direction opposite to that of the primary caloric response. We consider a true premature caloric reversal to be pathologic and indicative of CNS disease if the eardrums are intact (see Air calorics, pp. 180 to 183), if the secondary nystagmus begins prior to 140 sec after the onset of irrigation, and if slow-phase eye speed is more than 6° to 7°/sec.

Caloric inversion and perversion

The term caloric *inversion* refers to an entire caloric response that beats in the direction opposite to that expected. The term caloric *perversion* refers to the oc-

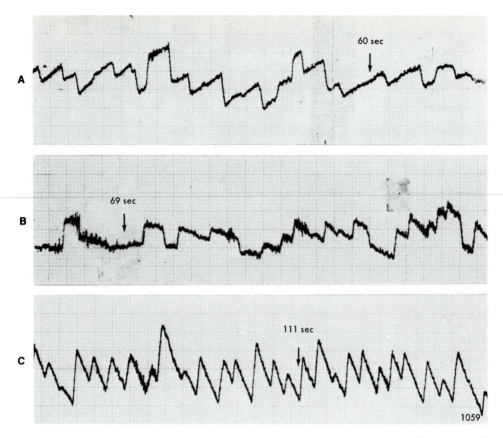

Fig. 9-9. Premature caloric reversal. The stimulus is a warm temperature irrigation of the right ear. **A,** A nystagmus response in the appropriate direction is present. **B,** The response subsides and secondary nystagmus begins. **C,** The secondary response continues to beat strongly. **A** to **C,** Bitemporal leads.

currence of vertical or oblique nystagmus as the response to a caloric irrigation.

Both caloric inversion and perversion are taken as evidence of brainstem disease. Cohen and Uemura[21] were able to produce perverted caloric nystagmus in monkeys both by electrolytic injury to the superior vestibular nucleus of one side and by unilateral injury to the rostral portion of the medial vestibular nucleus. Fredrickson and Fernandez[17] noted that cerebellovestibular lesions in cats also caused perverted caloric nystagmus.

Fig. 9-10 shows the peak caloric response from stimulation of the right ear with water at 20° C in a patient with right-sided brainstem infarction. The response to a lateral canal stimulus is vertical nystagmus beating upward. Horizontal nystagmus is absent entirely. This appears to be a true caloric perversion; the slow component velocity of the vertical nystagmus was somewhat less active earlier and later in the response. Conventional horizontal nystagmus, without vertical nystagmus, was obtained from right 44° and left 20° C irrigations.

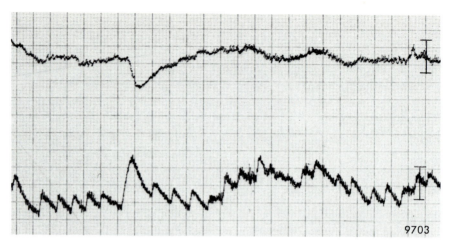

9703

Fig. 9-10. Perverted caloric response in patient with right-sided brainstem infarction. The stimulus is 20° C irrigation of right ear. Peak response shown. Top tracing, bitemporal horizontal leads; lower tracing, vertical leads, right eye. See text.

PITFALLS
Weak or absent response

There are times when one or more caloric irrigations provoke no response or one much weaker than expected. This event is always significant, and it requires further investigation to exclude errors. There are a number of possible causes for a weak or absent response, and each must be eliminated in turn. If a single response is weaker than the other three, it is likely either that the patient has a combination of unilateral weakness and directional preponderance or that the irrigation in question was inadequate. Once can distinguish between the two possibilities by examining the responses of the opposite ear. If the patient has a unilateral weakness and a directional preponderance, the effects of the preponderance are also seen in the opposite ear. As illustrated in Fig. 9-4, the patient had no left-beating response when the left ear was irrigated with warm water but a definite right-beating response when the same ear was irrigated with cool water. However, in the opposite ear, the right-beating response was also stronger than the left-beating response, so the lack of response following warm water irrigation of the left ear can be explained by a left weakness plus a right preponderance. Many combinations are possible, of course, depending on the relative severities of the unilateral weakness and directional preponderance, but the pattern of response strengths is always the same. If this pattern does not exist, the lack of response during one irrigation must be caused by inadequate irrigation, as shown in Fig. 9-11. In this case, no response was obtained when the left ear was irrigated with warm water, but the intensities of the other three responses are approximately equal; therefore, the absent response was probably caused by an inadequate irrigation.

The adequacy of irrigation can often be determined by otoscopic examination

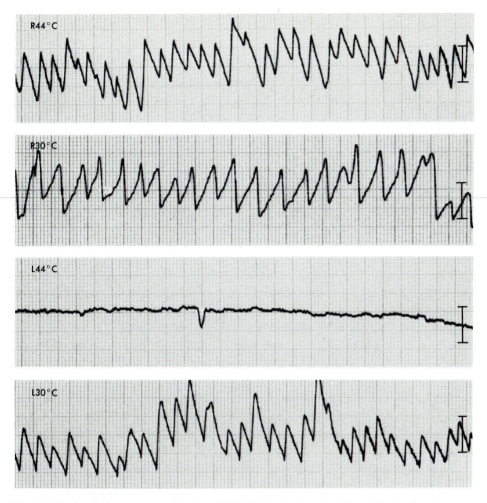

Fig. 9-11. Peak caloric responses of a normal individual, showing the effects of an inadequate warm temperature irrigation of the left ear. Bitemporal leads.

after irrigation. If the irrigation was adequate, a drop of water can often be seen resting against the tympanic membrane, and the membrane may appear more opaque than it was before the irrigation. When warm water is used, an adequate irrigation always causes pronounced vasodilation of the vessels along the handle of the malleus and the annulus of the tympanic membrane. Furthermore, if the external auditory canal is narrow and tortuous or filled with cerumen, hair, or other debris, there is increased probability that an inadequate irrigation is the cause of a weak or absent response. It is sometimes possible for one to determine whether a suspect irrigation was adequate by asking the patient if it sounded as loud to him as the others. A good irrigation sounds distinctly louder than a poor one, and this is useful information, provided of course that the patient has normal hearing in both ears. If an inadequate irrigation is suspected, then it must be repeated.

Sometimes none of the irrigations produces an adequate response. In this case, one must consider, in addition to inadequate irrigations, other possible sources of error.

One must be certain that the patient has been kept alert. Lack of alertness suppresses caloric nystagmus just as it does positional and gaze nystagmus. If the patient is taking CNS-depressant medications, it will be more difficult and perhaps impossible to alert him sufficiently, and one may have to ask him to return for a retest after discontinuing these medications.

One must also ensure that the lack of response with eyes closed is not caused by failure of fixation suppression; that is, one must observe that nystagmus is also absent when the patient's eyes are open.

Finally, one should ensure that the lack of response with eyes closed is not caused by a defect of saccadic eye movement. Such a defect would have been observed during the saccade test. To determine if an apparent lack of caloric response is caused by a saccadic defect, one must ask the patient to open his eyes when the peak caloric response is expected (between 45 and 90 sec after the beginning of the irrigation) and observe his eye position. If the eyes are deviated in the direction of the expected slow phase, a caloric response is indeed present, and no nystagmus is seen because the saccadic component is missing. This is a rare finding, but it should be considered when no response is elicited from either ear.

If both irrigations of a given ear provoke no response at all, the otoneurologist may wish to know whether vestibular function in that ear is totally lost or whether some residual function remains. To determine this, one can irrigate that ear with very cold water, such as 10° C water from the drinking fountain or even ice water, which produces a stronger stimulus than the standard irrigations and may provoke nystagmus where none was seen before. To perform the ice water caloric test, the technician draws 2 ml of the cold water into a syringe, then asks the patient to turn his head so that the ear to be irrigated is facing upward. The patient closes his eyes, the recorder is started, and the 2 ml of water are placed into the patient's external auditory canal. The water is allowed to remain in the canal for 20 sec, then the patient's head is turned to the opposite side so that the water runs out. The patient's head is then returned to the caloric test position. The ice water test may provoke some nystagmus beating toward the opposite ear where none was seen before, indicating that the vestibular function of that ear may be severely depressed but not entirely absent.

Great care must be taken in interpreting the nystagmus observed after the ice water caloric irrigation. Having ice water placed in his ear is extremely alerting to the patient, and the nystagmus the technician observes may be preexisting nystagmus, even if none had been seen before. To differentiate release of preexisting nystagmus from a true caloric response, one must carefully examine the record *immediately after* the water is placed in the ear. Any nystagmus seen at this point must be preexisting nystagmus; unless some abnormality of the external ear exists, the caloric response would have begun no sooner than 10 sec after introduction of the water.

Recall that the caloric irrigation affects only the lateral (and to a minor degree, superior) semicircular canal and central connections. If 2 ml of ice water provokes no nystagmus with eyes closed, one can be reasonably certain that the lateral canal system at least is nonfunctioning. However, function might still be retained in the vertical canal system, particularly the posterior canal; the caloric irrigation does not affect the posterior canal. Theoretically at least, residual dizziness after a disease or injury that causes loss of caloric (lateral canal system) function could be caused by abnormality in the posterior canal system.

Superimposed nystagmus

If nystagmus is present when the eyes are closed and the head is in the caloric test position, it summates algebraically with the caloric responses. The result is a directional preponderance in the direction of the preexisting nystagmus, which should be no cause for confusion. However, when the preexisting nystagmus is strong or other abnormalities are simultaneously present, a puzzling pattern of caloric responses can appear. Two examples illustrate this point.

The first example, shown in Fig. 9-12, is taken from a patient with vestibular neuronitis on the right side. He has preexisting left-beating nystagmus (Fig. 9-12, *A*). After irrigation of the (normal) left ear with cool water, faint right-beating nystagmus appears at the expected time of the peak caloric response (Fig. 9-12, *B*). About 30 sec later, left-beating nystagmus reappears (Fig. 9-12, *C*) and rapidly regains its pretest strength. The nystagmus reversal seen in Fig. 9-12, *C*, should not be misinterpreted as a premature caloric reversal (p. 172). It is merely the

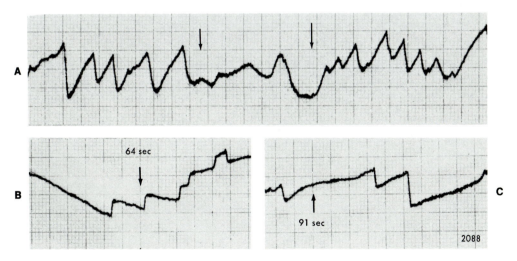

Fig. 9-12. Superimposed nystagmus. **A,** Preexisting left-beating nystagmus. There is a period of nystagmus suppression (between arrows) due to inadequate alerting. **B,** Appearance of the right-beating caloric nystagmus response. **C,** Resumption of preexisting nystagmus. A to C, Bitemporal leads.

reappearance of preexisting nystagmus that had been temporarily reversed by the caloric response.

The second example, shown in Fig. 9-13, is taken from the same patient. Pre-test nystagmus is still very strong (Fig. 9-13, *A*), and when his (abnormal) right ear is irrigated with warm water, the caloric response is not strong enough to over-come the preexisting nystagmus (Fig. 9-13, *B*). At the time of the expected peak caloric response, the preexisting nystagmus is weakened but not reversed in direction. In this case, in calculating maximum slow-phase eye speed of this response, one chooses the three beats with the *slowest* speeds and enters their mean in the formula as a negative value (since they are beating in the wrong direction). This response should not be misinterpreted as caloric inversion (pp. 172-173). The caloric response is in the appropriate direction; it is merely too weak to reverse the patient's preexisting nystagmus.

One should also be aware that some patients, despite alerting procedures, are alerted only when the irrigation starts. Fig. 9-14 illustrates this occurrence. When the patient's left ear is being irrigated with warm water, right-beating nystagmus appears. However, this is not a caloric inversion; in fact, it is not a caloric re-sponse at all. It begins immediately and persists unchanged in intensity. A caloric response would not have begun until about 20 sec after the beginning of the irriga-tion, and it would have shown a characteristic rise and gradual decline in intensity. It became evident that this patient had complete loss of vestibular response on the left side and (when sufficiently alerted) had a right-beating nystagmus with eyes closed.

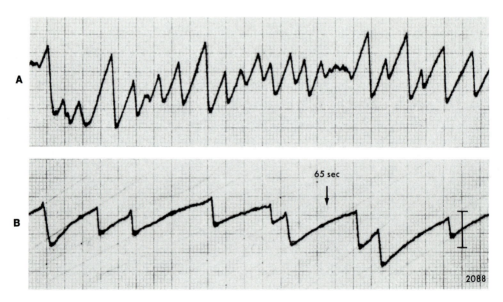

Fig. 9-13. Superimposed nystagmus. **A,** Preexisting left-beating nystagmus. **B,** Weakening, but not re-versal, of preexisting nystagmus by the caloric response. **A** and **B,** Bitemporal leads.

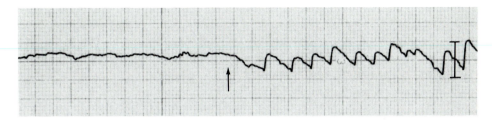

Fig. 9-14. Appearance of right-beating nystagmus immediately after the beginning of the irrigation (arrow). Bitemporal leads.

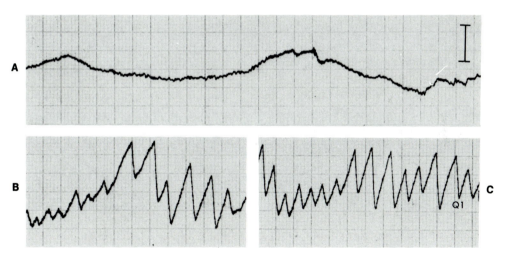

Fig. 9-15. Caloric response of patient with congenital nystagmus. A, Nystagmus absent with eyes closed. B, 30 sec after beginning of irrigation, nystagmus appears in appropriate direction. C, 60 sec after beginning of irrigation, nystagmus reaches peak intensity. A to C, Bitemporal leads.

Superimposed congenital nystagmus

Estimation of caloric function of the presence of congenital nystagmus is sometimes easy but often difficult and occasionally impossible. If the nystagmus is absent when the eyes are closed, standard estimations of the caloric response may be applied. This task is made much more difficult or impossible when (sometimes varying) nystagmus is present when the eyes are closed.

Fig. 9-15 shows the caloric response of a patient with congenital nystagmus that was abolished by eye closure (Fig. 9-15, *A*). A warm caloric stimulus was applied to the left ear. At about 30 sec after the irrigation begins, appropriate left-beating nystagmus appears and increases in speed (Fig. 9-15, *B*). The peak response occurs 60 sec after the beginning of the irrigation, and it then declines in the usual way (Fig. 9-15, *C*). The responses to the other caloric irrigations were similarly easy to read and quantify. However, fixation suppression of caloric nystagmus was difficult to estimate.

Fig. 9-16 shows the caloric response of a patient whose right-beating congenital nystagmus is still present with eyes closed (Fig. 9-16, *A*). Note the nonlinear form of the slow beat and the rounding of the tops of the beats. The response to cool caloric irrigation of the right ear at an expected time of maximum response is shown in Fig. 9-16, *B*. The direction of nystagmus is appropriate; and though there is still distortion of many beats, a number are "straight" enough for calculation of slow-component velocity. Calculation is more difficult for the same period of maximum response to cool caloric irrigation of the left ear (Fig. 9-16, *C*); the beat distortion is quite uniform, but the speed of the slow component is visibly greater than that in Fig. 9-16, *B*, which is to be expected in the presence of right-beating nystagmus before stimulus.

In some records, the response to caloric stimulus in patients with congenital nystagmus is so uncertain as to make the record indecipherable, although marked directional preponderance is not uncommon. In such cases, the identification of a response depends on consideration of such factors as expected latency, time course, and direction of nystagmus; even then, little more than an informed guess can be made regarding the caloric activity.

Air calorics

Warm and cool air caloric stimulation ordinarily produces nystagmus responses that are quite similar to those produced by water stimulation (Table 9-2). However, in the presence of a medium or large perforation of the tympanic membrane, the response to warm air stimulation frequently mimics the appearance of caloric inversion (pp. 172-173), except that reversal occurs a little later.

The example shown in Fig. 9-17 is taken from a patient who has a large posterior perforation of the right tympanic membrane but no CNS disease. No nystagmus is present before the test, when the patient's eyes are closed and effective alerting mechanisms are used (Fig. 9-17, *A*). After the warm air stimulus is applied to the right ear, nystagmus appears with latency of only 5 sec, beating toward the

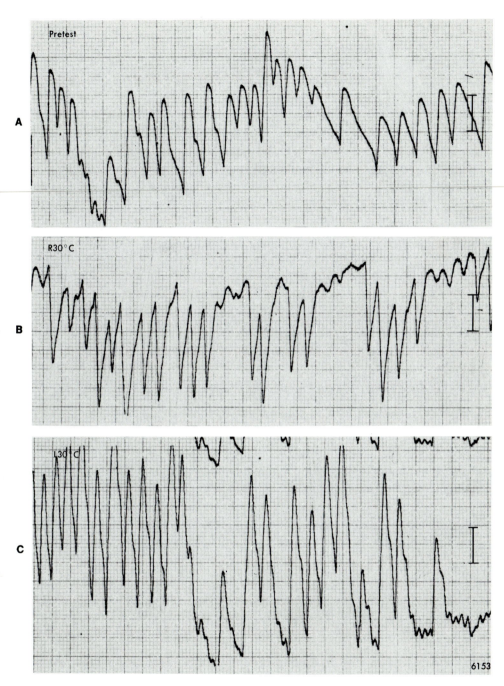

Fig. 9-16. Caloric response of patient with congenital nystagmus. **A,** Right-beating nystagmus present with eyes closed. **B,** Peak response to cool irrigation of right ear. **C,** Peak response to cool irrigation of left ear. **A** to **C,** Bitemporal leads.

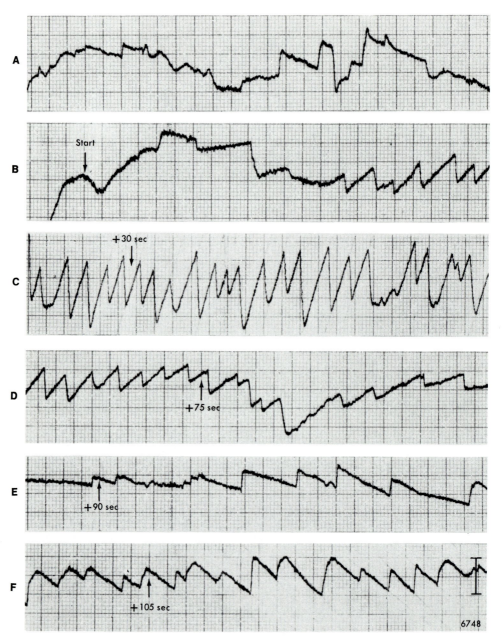

Fig. 9-17. Caloric response of patient with large perforation of right tympanic membrane. Stimulus is warm air to right ear. **A,** No preexisting nystagmus with eyes closed. **B,** Inappropriate left-beating nystagmus begins about 5 sec after stimulus is applied. **C,** Left-beating nystagmus reaches peak intensity after 30 sec. **D,** Left-beating nystagmus declines after 75 sec. **E,** Appropriate right-bearing nystagmus appears after 90 sec. **F,** Right-beating nystagmus gains intensity. A to F, Bitemporal leads.

left, opposite the expected direction (Fig. 9-17, *B*). The left-beating response reaches its maximum at about 30 sec (Fig. 9-17, *C*), then declines in intensity after 75 sec (Fig. 9-17, *D*). There is a period of about 5 sec between Fig. 9-17, *D* and *E*, in which nystagmus is absent. At about 90 sec, right-beating nystagmus appears (Fig. 9-17, *E*), gains intensity, and soon reaches its maximum speed (Fig. 9-17, *F*).

Important features of this unusual response are its short latency from stimulus, the unexpected direction of the first response, and the appropriate direction of the secondary response. The phenomenon occurs only with warm air stimulation, never with cool. Barber and co-workers[22] have shown that it results from evaporative cooling of the damp middle ear mucous membrane in the presence of tympanic membrane perforation or mastoidectomy cavity, or even in a normal ear with moisture in its depths.

If air caloric tests are used, it is important to be aware of this occurrence in order to avoid ascribing its effects to brainstem disease.

VERTICAL LEAD IN THE CALORIC TEST

In caloric testing with eyes closed, pen movements similar or identical to those of jerk nystagmus often appear in the vertical lead. These might result from blink,[23] which at times closely mimics ordinary jerk nystagmus, electronic artifact from cross-interference between horizontal and vertical channels, oblique nystagmus from stimulus, or a combination of these. The suggested procedure for decision making on this matter is as follows:

1. If the vertical "nystagmus" has a sharp peak at the apex of a curved slow beat (Fig. 9-18), it is blink.

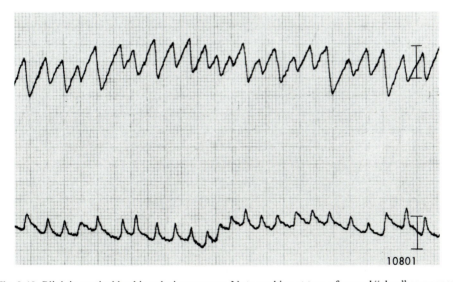

Fig. 9-18. Blink in vertical lead in caloric response. Note peaking at tops of curved "slow" component of beats. Peak response, stimulus 30° C in right ear. Top tracing, bitemporal horizontal leads; lower tracing, bitemporal vertical leads.

2. If the eye movements have the wave form of typical jerk nystagmus (Fig. 9-19), first check the calibration record for possible cross-interference between horizontal and vertical channels. Horizontal calibration movements may cause significant pen deflection in the vertical channel. Note the directions of pen movement in the two channels. In Fig. 9-19, *B*, 20° horizontal eye movement to the right causes about 7° vertical movement upward, and this relationship precisely explains the vertical "nystagmus" in Fig. 9-19, *A*.

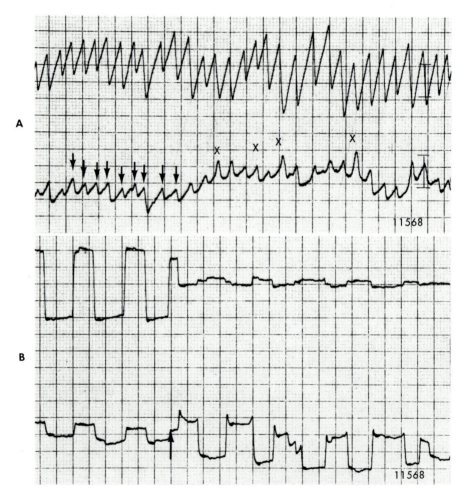

Fig. 9-19. ? Nystagmus in vertical lead. **A,** Peak caloric response, stimulus 44° C in left ear. In vertical lead, beats shown by arrows appear to be typical jerk nystagmus. Beats marked *X* are certainly typical blink. **B,** Horizontal (left of arrow) and vertical (right of arrow) 20° calibrations. As the eyes move horizontally 20° to *right,* there is pen displacement about 7° *upward* in the vertical lead. This cross-interference explains the apparent downward beating vertical "nystagmus" in **A. A** and **B,** Top tracing, bitemporal horizontal leads; lower tracing, vertical lead, right eye.

3. If in the calibration record the vertical pen displacement caused by horizontal eye movement does not correspond to the direction of "nystagmus" in the vertical lead during caloric response, one assumes that the cross-interference is irrelevant and that there is actually a vertical component to the nystagmus. Fig. 9-20 illustrates this occurrence.

Combinations of blink and nystagmus probably occur as well, and decision making is then difficult. Vertical channel pen movements are so common in the

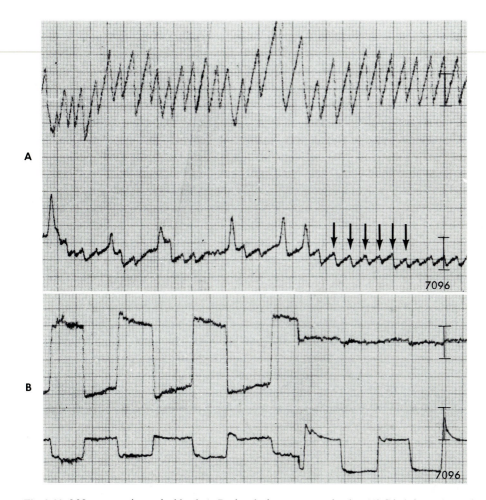

Fig. 9-20. ? Nystagmus in vertical lead. **A,** Peak caloric response, stimulus 44° C in left ear. In vertical lead, beats shown by arrows appear to be typical jerk nystagmus. **B,** horizontal (left of arrow) and vertical (right of arrow) 20° calibrations. As the eyes move horizontally 20° to *right,* there is pen displacement about 7° *downward* in the vertical lead. In **A,** note that *left-beating* nystagmus in the horizontal lead corresponds to *downward* beating vertical nystagmus. The "nystagmus" in the vertical lead therefore cannot be due to cross-interference between horizontal and vertical. One assumes it represents a true eye movement, the quick component of the nystagmic response beating obliquely to left and downward. **A** and **B,** Top tracing bitemporal horizontal leads; lower tracing, vertical lead, right eye.

Table 9-3. Summary of abnormalities observed in the caloric test

Abnormality	Significance	Comment
Unilateral weakness (p. 161)	Usually peripheral vestibular lesion (weak ear)	Rule out inadequate irrigation(s), lack of alertness
Directional preponderance (pp. 161-162)	Peripheral vestibular lesion or CNS lesion	
Bilateral weakness (pp. 165-169)	Bilateral peripheral vestibular lesion or CNS lesion	Rule out inadequate irrigation(s), CNS-depressant drugs, lack of alertness, failure of fixation suppression, saccadic defect
Hyperactive response (p. 170)	Overalertness, CNS lesion	
Failure of fixation suppression (pp. 170-172)	CNS lesion	Rule out drugs
Premature caloric reversal (p. 172) Caloric inversion (pp. 172-173)	CNS lesion	Rule out superimposed preexisting or congenital nystagmus, air caloric with tympanic membrane perforation, wet ear canal
Caloric perversion (pp. 172-173)	CNS lesion	

eyes-closed caloric response that we tend to ignore them at present, unless regular jerk nystagmus appears that cannot be explained by a cross-interference effect. True caloric perversion (p. 174, Fig. 9-10) is a very rare event and should be diagnosed only with great caution.

REFERENCES

1. Capps, M. J., Preciado, M. C., Paparella, M., and Hoppe, W. E.: Evaluation of air caloric tests as a routine examination procedure, Laryngoscope 83:1013, 1973.
2. Ford, C. R., and Stockwell, C. W.: Reliabilities of air and water caloric responses, Arch. Otolaryngol. 104:380, 1978.
3. Barber, H. O., and Wright, G.: Unpublished data.
4. Baloh, R. W., and Honrubia, V.: Clinical neurophysiology of the vestibular system, Philadelphia, 1979, F. A. Davis Co., p. 136.
5. Hamersma, H.: The caloric test: a nystagmographical study, doctoral thesis, University of Amsterdam, Amsterdam, 1957, N.V. Drukerii Gebr. inten Bergen op Zoom.
6. Jongkees, L. B. W., and Philipszoon, A. J.: Electronystagmography, Acta Otolaryngol., Supp. 189, 1964.
7. Mehra, Y. G.: Electronystagmography: a study of caloric tests in normal subjects, J. Laryngol. Otol. 78:520, 1964.
8. Henriksson, N. G.: Speed of slow component and duration in caloric nystagmus, Acta Orolaryngol., Supp. 125, 1956.
9. Benitez, J. T., Bouchard, K. R., and Choe, Y. K.: Air calorics: a technique and results, Ann. Otol. Rhinol. Laryngol. 87:216, 1978.
10. Alpert, J. N.: Failure of fixation suppression: a pathologic effect of vision on caloric nystagmus, Neurology 24:891, 1974.
11. Coats, A. C.: Directional preponderance and unilateral weakness as observed in the electronystagmographic examination, Ann. Otol. Rhinol. Laryngol. 74:655, 1965.
12. Cogan, D. G.: Syndrome of nonsyphilitic interstitial keratitis and vestibuloauditory symptoms, Arch. Ophthalmol. 33:144, 1945.
13. Smith, J. L.: Cogan's syndrome, Laryngoscope 80:121, 1970.
14. Simmons, F. B.: Patients with bilateral loss of caloric response, Ann. Otol. Rhinol. Laryngol. 82:175, 1973.
15. Stroud, M. H.: The otologist and the midline cerebellar syndrome, Laryngoscope 77:1795, 1967.

16. Torok, N.: The hyperactive vestibular response, Acta Otolaryngol. **70:**153, 1970.
17. Fredrickson, J. M., and Fernandez, C.: Vestibular disorders in fourth ventricle lesions, Arch. Otolaryngol. **80:**521, 1964.
18. Coats, A. C.: Central electronystagmographic abnormalities, Arch. Otolaryngol. **92:**43, 1970.
19. Takemori, S.: Visual suppression of vestibular nystagmus after cerebellar lesions, Ann. Otol. Rhinol. Laryngol. **84:**318, 1975.
20. Milojevic, B., and Allen, T.: Secondary phase nystagmus: the caloric test, Laryngoscope **77:**187, 1966.
21. Cohen, B., and Uemura, T.: Ocular changes in monkeys after lesions of the superior and medial vestibular nuclei and the vestibular nerve roots. In Naunton, R. F., editor: The vestibular system, New York, 1975, Academic Press, Inc., p. 187.
22. Barber, H. O., Harmand, W. M., and Money, K. E.: Air caloric stimulation with tympanic membrane perforation, Laryngoscope **88:**11-17, 1978.
23. Barry, W., and Melvill-Jones, G.: Influence of eye lid movement upon electro-oculographic recording of vertical eye movements, Aerospace Med. **36:**855, 1965.

Chapter 10

Evaluating and reporting the findings

Most ENG examinations are made by technicians. The results are sometimes analyzed and reported by an otoneurologist, but more often the technician reports the results directly to a referring otolaryngologist or neurologist. The analytic philosophy or outlook presented in this chapter applies especially to the second circumstance.

EVALUATING THE FINDINGS
Clinical importance of ENG

The role of ENG in vestibular diagnosis must be kept within a reasonable perspective. Essential diagnostic information is obtained by the physician from a number of other sources, such as history, conventional otoneurologic physical examination, audiologic data, and appropriate radiologic examination. To an experienced otoneurologist, the patient's history alone provides a reasonable diagnosis in many cases.

It is true that the ENG examination sometimes gives critical information. For example, gaze nystagmus might be revealed only on ENG examination, or ageotropic positional nystagmus might be recorded only when the patient's eyes are closed. These instances in which the ENG examination has unique value are quite unusual, however. In most cases, if a marginal ENG abnormality is overlooked, the patient will probably come to no serious harm, because a number of other diagnostic approaches will be brought to bear on his problem.

The ENG report should be viewed by the informed physician as confirmatory, as serving to identify unexpected but important diagnostic features (such as congenital nystagmus or bilateral caloric loss), as excluding certain differential diagnostic considerations, or as irrelevant to the clinical problem even if of academic interest. The physician hopes that the findings from this noninvasive investigation will spare his patient more hazardous diagnostic studies. Sometimes it does; sometimes it does not.

Quality of the tracing

In evaluating the tracing, the technician first scans it rapidly, looking for qualitative features that may influence the reliability of the interpretation and at the same time acquiring a "feel" for the patient as a bioelectrical entity. The technician notes whether the tracing is clean or noisy and whether the baseline is stable or shifting excessively. The record may be featured by sinusoidal oscillations and frequent suppression periods in the caloric nystagmus (generally indicating a low level of alertness) on the one hand, or by excessive quiver or microsaccadic movement, muscle potentials, blinking, or square wave movements (indicating a nervous patient) on the other. Noisy records add greatly to one's difficulty of interpretation, especially in measuring slow-phase eye speeds of low-intensity nystagmus beats. Clean, stable records add to one's confidence in reliability of the interpretation.

Weighting of findings

It is advisable for the physician to base the analysis more on definite, or "hard," findings than on marginal, or "soft," findings that may be of interest mainly to the technician. (It is true, of course, that some of today's marginal findings may be significant tomorrow, but the reverse is also true.) For example, a significant caloric reduction based on reliable mathematical values is a "hard" finding of an abnormality, suggesting peripheral localization and (bi)laterality as well. Directional preponderance of caloric nystagmus as an isolated finding, on the other hand, is usually interpreted as a "soft" abnormality, indicative of a possible vestibular system lesion of unknown localization.

In addition, one must consider the fact that many tests, even the quantitative caloric test, have a broad range of normal values, poorly defined normal values, or such variability in clinical application of test stimuli that other laboratories' normal values may not apply. Ideally, each ENG laboratory should establish its own range of normal values with well-controlled standardized stimuli, but this state of perfection will probably never be attained. Thus, conservative interpretation of test results, which is likely to enhance the physician's confidence in the value of an ENG report, is of great importance.

Admittedly, it is sometimes difficult to tread a sure path when differentiating between a minor alteration that is insignificant and a minor, subtle change that denotes disease. As an example, it may be impossible to decide whether a very minor degree of saccadic alteration of smooth pursuit is a normal variation or represents brainstem disease. The decision would be made through consideration of other findings as well (see Table 10-1). Conservative interpretation is advisable, because patients are less likely to come to harm (or to expensive, even hazardous, investigation) from an underanalyzed than from an overanalyzed ENG record.

Varying pathologic conditions

It is important to recall that ocular signs, including horizontal or vertical nystagmus, may vary from time to time as the underlying pathologic condition varies. Neoplastic, ischemic, or drug-induced conditions located within the vestibular system are prone to change. It is wrong to assume that the harmful effects of tu-

Table 10-1. Reliability and localizing value of ENG findings

Test	Abnormality	Reliability[a]
Gaze	Horizontal nystagmus (follows Alexander's law)	+++
	Bilateral gaze nystagmus, eyes open	+++
	Bilateral gaze nystagmus, eyes closed	++
	Unilateral gaze nystagmus, eyes open	+++
	Unilateral gaze nystagmus, eyes closed	+
	Rebound nystagmus	+++
	Periodic alternating nystagmus	+++
	Upbeating nystagmus	+++
	Downbeating nystagmus	+++
	Pendular nystagmus	+++
	Square wave movements	++
	Internuclear ophthalmoplegia	+++
Saccade	Ocular dysmetria	+++
	Saccadic slowing	+++
	Internuclear ophthalmoplegia	+++
Tracking	Saccadic pursuit	+++
	Disorganized pursuit	+++
	Disconjugate pursuit	+++
Optokinetic	Asymmetry	+++
	Declining response to increasing stimulus speeds	+++
	Inversion	+++
Positional	Direction-fixed nystagmus, eyes open	+++
	Direction-changing nystagmus, eyes open	+++
	Direction-fixed nystagmus, eyes closed	++
	Direction-changing nystagmus, eyes closed	++
	Direction-changing nystagmus, single head position	+++
Hallpike maneuver	Unilateral benign paroxysmal type positioning nystagmus	+++
	Bilateral benign paroxysmal type positioning nystagmus	+++
	Any other nystagmus	++
Caloric	Unilateral weakness	+++
	Directional preponderance	+
	Bilateral weakness	+++
	Hyperactive response	+
	Failure of fixation suppression	+++
	Premature caloric reversal	++
	Caloric inversion	+++
	Caloric perversion	+++

[a]+++ A "hard" finding, nearly always denotes a lesion; ++ an "intermediate" finding, usually denotes a lesion; + a "soft" finding, sometimes denotes a lesion.

Localization[b]		Comment
Peripheral vestibular	CNS	
+++	0	
0	+++	
0	+++	
+	++	
++	+	
0	+++	Cerebellar system lesion
0	+++	Usually posterior fossa lesion
0	+++	Lesion, drug induced
0	+++	Lower medullary lesion
0	+++	Usually congenital nystagmus
0	+++	Overalert patient, occasionally cerebellar system lesion
0	+++	Medial longitudinal fasciculus lesion
0	+++	Cerebellar system lesion
0	+++	Saccadic system (supranuclear) lesion
0	+++	Medial longitudinal fasciculus lesion
0	+++	
0	+++	
0	+++	
0	+++	Supratentorial, brainstem lesion
0	+++	Usually brainstem lesion
0	+++	Usually congenital nystagmus
0	+++	
0	+++	
++	+	
+	++	If ageotropic, probably CNS lesion (except PAN II)
0	+++	
++	+	Usually undermost ear lesion
++	++	Both ears, CNS lesion
+	++	
+++	+	Almost always weak ear lesion
+	++	Localization uncertain
++	+	Both ears, CNS lesion
0	+++	Overalert patient, cerebellovestibular disease
0	+++	
0	+++	
0	+++	Brainstem lesion
0	+++	Brainstem lesion

[b]+++ Nearly always denotes the indicated site of lesion; ++ usually denotes the indicated site of lesion; + sometimes denotes the indicated site of lesion; 0 almost never denotes the indicated site of lesion.

mor, for example, are fixed and immutable. They are not; considerable variability from time to time in the lesion and in its physical signs is usual. Thus, nystagmus might be unequivocal one day, absent the next, and marginal on another. This adds to the problem of ENG analysis, because (though this is not the rule) major lesions involving the vestibular system at times cause minor or marginal ENG abnormalities, or none at all. Multiple sclerosis is a familiar example.

Reviewing the entire record

The ENG record must be interpreted as a whole. As in audiology, if multiple abnormal features are present, one can identify and localize the abnormality with greater assurance. When the patient has CNS disease, it is almost the rule that more than one clear ENG defect will be found, for example, bilateral gaze nystagmus with defective smooth pursuit and saccadic overshoots. Single findings in this circumstance, unless unequivocal, should give pause to a confident interpretation of CNS localization. Peripheral disorders may be featured by caloric reduction as an isolated finding, but even in this instance, significant positional nystagmus is quite common, and the absence of CNS abnormalities should be taken into account in the analysis.

Using multiple leads

Some ENG abnormalities may be identified only if monocular leads are used. Reference to this has been made in different sections of Chapters 5 and 6. Internuclear ophthalmoplegia, disconjugate eye movements, and identification of faint nystagmus in the abducting eye (when there is doubt as to whether nystagmus is present) are examples. The advantages of using monocular leads routinely should not be overlooked if patients with neurologic disorders are referred to the ENG laboratory. A single vertical lead also permits identification of blinks, their possible role in saccadic overshoot, blink distortion of the bitemporal lead, vertical nystagmus, and vertical gaze palsy.

If the ENG interpreter has only a single (bitemporal) lead on which to base judgment, he should recognize the limitations of his technical method and understand that certain findings may not be detected.

Identifying specific abnormalities

The examiner should look carefully for those abnormalities, admittedly infrequent in the usual referral population, that carry considerable diagnostic specificity. A partial list includes:

1. A great deal of pendular or jerk nystagmus, distorted beat form, difficulty with caloric calculation, directional preponderance in caloric calculation, which denote congenital nystagmus

2. Saccadic overshoots and undershoots, which denote cerebellar system disease

3. Failure of fixation suppression, which denotes CNS disease, drug intoxication, or metabolic effect

4. Bilateral, horizontal gaze nystagmus with eyes open, or oblique or vertical beating nystagmus, which denotes CNS disease or possible drug intoxication

5. Rebound nystagmus, which denotes cerebellar system disease

6. Vertical downbeating nystagmus in lateral gaze with eyes open, which denotes CNS disease, medullocervical localization

7. Positioning nystagmus in a single Hallpike position, bitemporal lead beating to "wrong" side, which usually denotes unilateral inner ear disease of undermost ear

8. Bilateral internuclear ophthalmoplegia, which usually denotes multiple sclerosis

Table 10-1 is a summary guideline for record analysis. The weightings on reliability and localization reflect the authors' current views and, of course, are subject to future modification. The findings should be recorded on a suitable laboratory work sheet, which should be retained as part of the patient's permanent record.

REPORTING THE FINDINGS

Analysis and reporting of the record require decision and comment on the following points:

1. Is the record normal (including normal variations) or abnormal?

2. If abnormal, is it possible to identify localization of the pathologic condition, peripheral vestibular versus CNS lesion?

3. If a peripheral lesion is detected, can it be lateralized? If a CNS lesion is detected, can it be further localized? (Answer: usually not.)

4. Can specific etiology of an abnormal recording be reasonably suggested? (Answer: rarely.)

Figs. 10-1 and 10-2 show examples of ENG reports similar to those sent to referring physicians from the Sunnybrook Medical Centre ENG Laboratory. The top portion incorporates a summary of some of the objective findings and is filled out in check-off form by the technician. The lower part contains analytical comments designed to give as much diagnostic information as is warranted by the findings. Our practice is to interpret the tracings with no knowledge of the patient's history, physical findings, audiologic examination results, or tentative diagnosis, so as to avoid bias from these sources.

The reader will see that the data summary portion of the work sheet omits many pertinent details of the examination. This is deliberate; the report will be read by physicians of greatly varying degrees of knowledge of the vestibular system, and the purpose of the report is to give clinical information that is as accurate and practical as possible rather than to confuse. Our practice is to select pertinent details from the tracings and mention only those in the report. Of course, full and detailed information on the examination is recorded on the technician's work sheet, which is retained in the laboratory.

Some workers include photocopy extracts of interesting portions of the tracings with the report of the examination. Whether this practice is followed or not, it

Sunnybrook Medical Centre

Dizziness Unit

Name: Age   Date:

Address: Sex: Referred by:

                               File: By:

All tests performed with electronystagmography

1. The caloric tests show:

☐ Normal responses each ear

☐ Responses apparently reduced bilaterally

☒ Minor significant reduction   *left*   ear

☐ Moderate significant reduction   ear

☐ Marked significant reduction   ear

☐ No response with ice water   sec., right ear ☐

                                          sec., left ear   ☐

2. Posture test shows signficant nystagmus is: Present ☒
                                                  Absent ☐

3. Other tests

                    ☒ Recorded   ☒ Recorded
    Gazes: Tracking
                    ☐ Not recorded   ☐ Not recorded

                    ☒ Recorded
    OKN:
                    ☐ Not recorded

Interpretation and comments: This interpretation is deliberately based on study of ENG
                              records alone.

Abnormal ENG.

Significant caloric reduction L (UWL 29%), left-beating positional nystagmus of
sufficient slow component velocity to be considered pathological. ENG stigmata
of CNS disease absent. Findings consistent with peripheral lesion left side.

Fig. 10-1. Sample ENG report.

Sunnybrook Medical Centre

Dizziness Unit

Name: Age   Date:

Address: Sex: Referred by:

                                File: By:

All tests performed with electronystagmography

1. The caloric tests show:

   ☒ Normal responses each ear

   ☐ Responses apparently reduced bilaterally

   ☐ Minor significant reduction   ear

   ☐ Moderate significant reduction   ear

   ☐ Marked significant reduction   ear

   ☐ No response with ice water   sec., right ear ☐

                                                 sec., left ear   ☐

2. Posture test shows signficant nystagmus is: Present ☐
                                                   Absent   ☒

3. Other tests

   Gazes: ────< ☒ Recorded   Tracking ────< ☒ Recorded
               ☐ Not recorded   ☐ Not recorded

   OKN: ────< ☒ Recorded
             ☐ Not recorded

Interpretation and comments: This interpretation is deliberately based on study of ENG
                             records alone.

Abnormal ENG.

There are repeated saccadic overshoots on calibration movements, saccadic pursuit, bilateral horizontal gaze nystagmus. Fixation suppression of caloric nystagmus is defective. Findings indicate well-established cerebellar system disease.

Fig. 10-2. Sample ENG report.

is essential to retain possession of the entire ENG record in the laboratory. The original tracings are the basic evidence and only enduring record of the study, and they are as important to the ENG examiner as are radiographs to the radiologist.

Effective medical use of a competent ENG examination and report presupposes some knowledge of vestibular disorders on the part of the referring physician. Unfortunately, some physicians naively expect a specific diagnosis to emerge from the study; they are almost invariably disappointed in this hope. The example shown in Fig. 10-1 gives clear evidence of a left-sided peripheral vestibular lesion but not its cause. The physician's responsibility is to assign this valuable information to its proper place in the diagnostic evaluation and to differentiate possible causes (such as Meniere's disease, labyrinthitis, acoustic neuroma, or poststapedectomy fistula) on other grounds.

A final word of caution. Knowledge of the ENG features that indicate vestibular-oculomotor-cerebellar system disorders is still in its infancy. Lesions (and hence such physical signs as nystagmus) change from day to day, and future research is certain to assign today's fact to tomorrow's fallacy, and vice versa. Thus, one must be conservative in the reporting of the ENG record, especially where the findings are "soft." For example, as a yardstick for clear identification of nystagmus, we believe that it is necessary to see three consecutive beats in a reasonable time period, each with a readily recognizable slow and quick component. One or even two such beats could conceivably represent random eye movement, especially with eyes closed. Good judgment, a major part of the armamentarium of the skilled ENG technician or otoneurologist, comes with increasing knowledge and experience.

Chapter 11

Illustrative cases

Having had an opportunity to inspect many ENG tracings illustrating a wide variety of abnormalities, the reader may now test his own skill at test interpretation. Some of the cases given below are straightforward, some are tricky, but the reader can correctly interpret all cases using only the information supplied. Answers to the questions begin on p. 220.

CASES

Case 1. The tracings in Fig. 11-1 are portions of a gaze record.

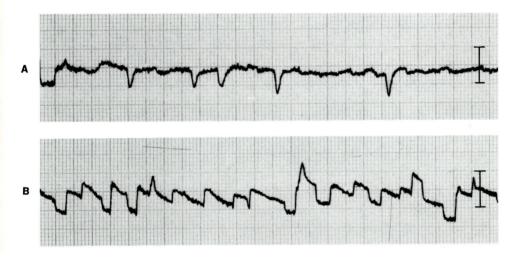

Fig. 11-1. **A,** Eyes open. **B,** Eyes closed. **A** and **B,** Center gaze, bitemporal leads.

QUESTION:

How would you interpret these tracings?

Case 2. The tracing in Fig. 11-2 is also a portion of a gaze record.

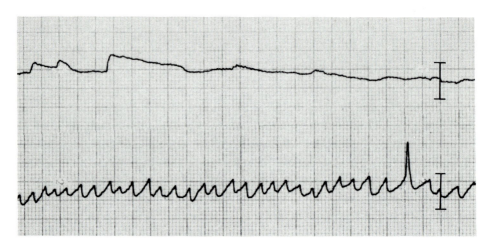

Fig. 11-2. Eyes open, center gaze. Upper tracing, bitemporal leads; lower tracing, vertical leads.

QUESTIONS:

a. How would you interpret this tracing?

b. Can the patient's lesion be localized on the basis of this tracing?

Case 3. The tracings in Fig. 11-3 are portions of the record obtained during the gaze, tracking, and caloric tests. No abnormalities other than those shown were detected.

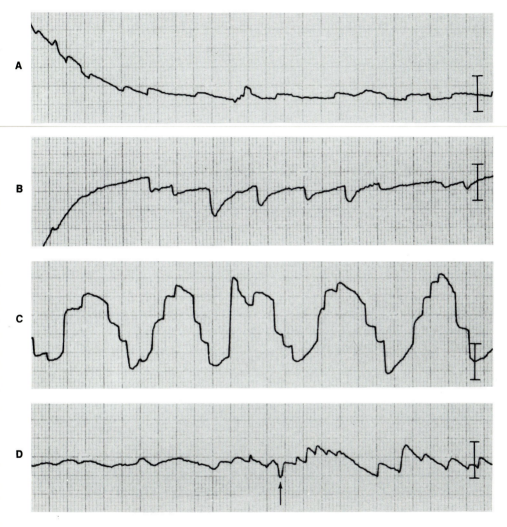

Fig. 11-3. **A,** Rightward gaze, eyes open. **B,** Leftward gaze, eyes open. **C,** Tracking. **D,** Caloric response to warm temperature irrigation of the right ear; eyes are closed to the left of arrow and open to the right of arrow. **A** to **D,** Bitemporal leads.

QUESTIONS:

a. What abnormalities are revealed by these tracings?
b. Can the patient's lesion be localized on the basis of these abnormalities?
c. What alternative explanations must be considered before one concludes that an organic lesion is responsible for these abnormalities?

Case 4. The tracings in Fig. 11-4 were taken from an epileptic patient. Caloric re-sponses were weak bilaterally, but not below normal limits. The remainder of the record showed no abnormalities.

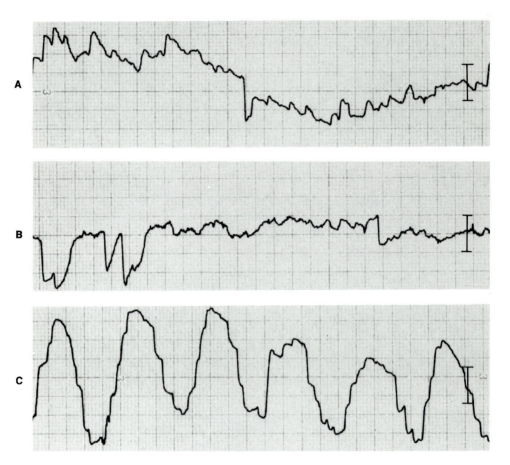

Fig. 11-4. A, Rightward gaze, eyes open. **B,** Leftward gaze, eyes open. **C,** Tracking, A to C, Bitemporal leads.

QUESTIONS:

a. What abnormalities are revealed by these tracings?

b. What is a likely explanation of these abnormalities?

Case 5. The tracings in Fig. 11-5 are portions of the record obtained during the positional test; the patient has positional vertigo.

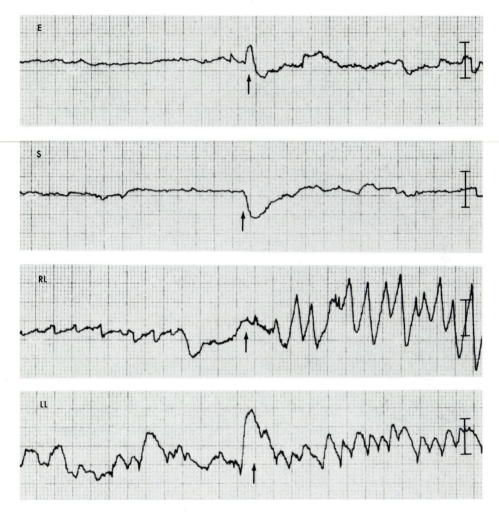

Fig. 11-5. All tracings: eyes open and fixating, center gaze to left of arrow, closed to right of arrow, bitemporal leads. In Figs. 11-5 and 11-6, *E* refers to erect position, *S* to supine, *RL* to right lateral, and *LL* to left lateral.

QUESTIONS:

a. What abnormality is revealed?

b. Does this abnormality help localize the patient's lesion?

Case 6. The tracings shown in Fig. 11-6 were also obtained during a positional test. Only the RL and LL positions are shown; no nystagmus was present in the other positions. The patient also had unilateral gaze nystagmus and saccadic pursuit eye movements.

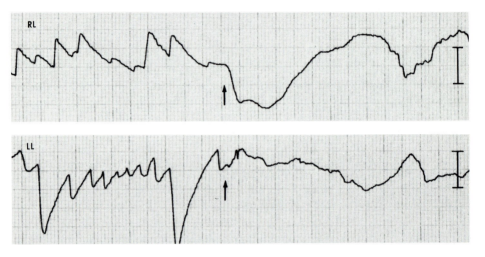

Fig. 11-6. Both tracings: eyes open to the left of arrow and closed to the right of arrow, bitemporal leads.

QUESTIONS:

a. What abnormality is revealed?

b. Does this abnormality help localize the patient's lesion?

Case 7. An ENG examination was performed on a patient who complained of transient vertigo when he lay down in bed. The referring physician reported that at first he had suspected benign paroxysmal vertigo. The Hallpike maneuver provoked transient vertigo in the HHR position, but it also provoked a nystagmus response that was not delayed in onset, not transient, and not fatigable; thus the physician suspects a CNS lesion. The ENG examination yielded the tracing shown in Fig. 11-7 while the patient was gazing straight ahead with eyes open.

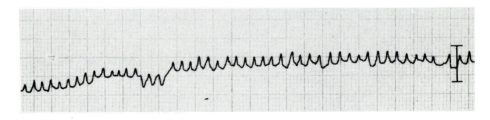

Fig. 11-7. Center gaze, eyes open, bitemporal leads.

QUESTION:

What is the probable explanation of the nystagmus response seen by the referring physician during the Hallpike maneuver?

Case 8. The tracings in Fig. 11-8 are segments showing the peak caloric responses for each of the four caloric irrigations.

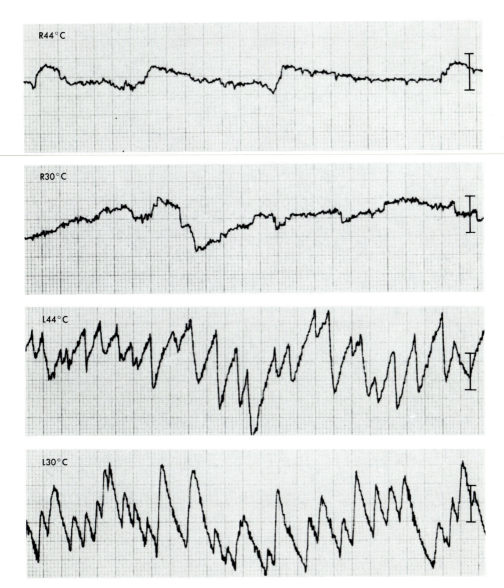

Fig. 11-8. Peak caloric responses, bitemporal leads.

QUESTIONS:

a. What abnormality is revealed?

b. What alternative explanations must be considered before one concludes that the abnormality is caused by an organic lesion?

Case 9. The tracings in Fig. 11-9 are also segments showing peak caloric responses for each of the irrigations.

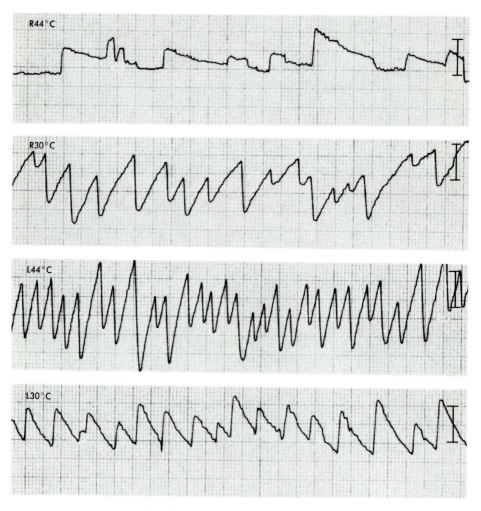

Fig. 11-9. Peak caloric responses, bitemporal leads.

QUESTION:

What abnormalities are revealed?

Case 10. The tracings in Fig. 11-10 are also segments showing peak caloric responses for each of the irrigations.

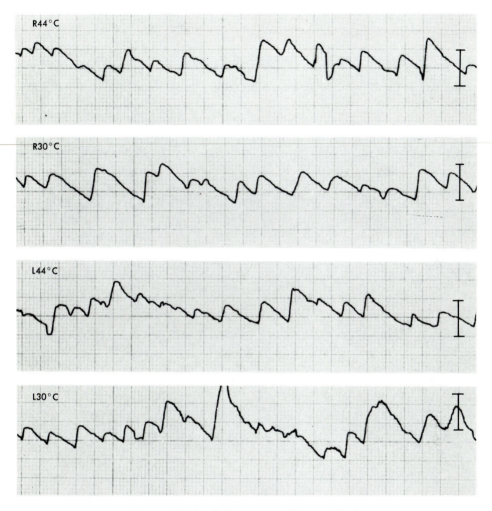

Fig. 11-10. Peak caloric responses, bitemporal leads.

QUESTION:
What abnormality is revealed?

Case 11. The tracings in Fig. 11-11 are also segments showing the peak responses for each of the irrigations.

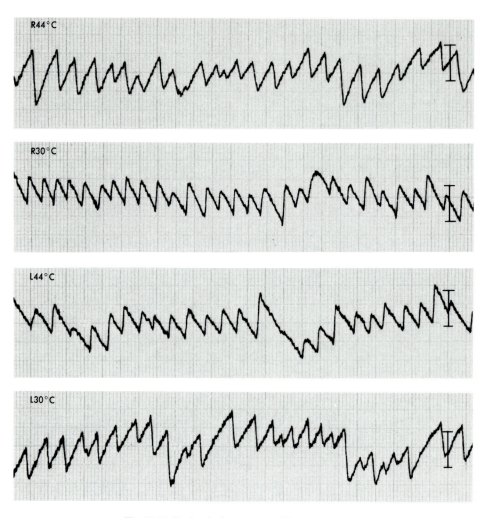

Fig. 11-11. Peak caloric responses, bitemporal leads.

What abnormality is revealed, and what is the most probable explanation?

Case 12. The tracing in Fig. 11-12 shows the effect of visual fixation on caloric nystagmus.

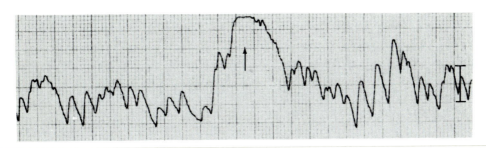

Fig. 11-12. Caloric response to a warm irrigation of the right ear. Eyes are closed to the left of arrow and open to the right of arrow, bitemporal leads.

QUESTION:

Does this tracing reveal an abnormality?

For each of the following cases, write a report on the ENG examination for the referring physician.

Case 13

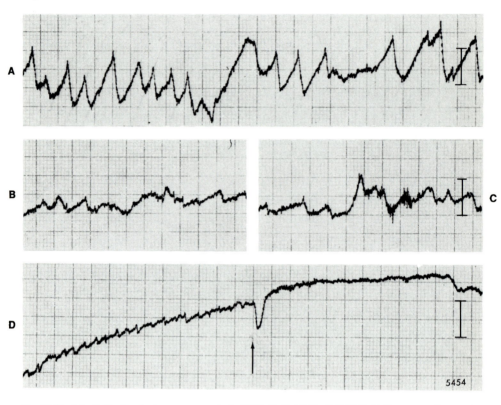

Fig. 11-13. A to C, Peak caloric responses: **A,** L 44° C; **B,** R 44° C; **C,** R 0° C; **D,** Gaze record, eyes deviated to left; eyes open to left of arrow and closed to right of arrow. All tracings, bitemporal leads. Saccade and tracking tests normal; nystagmus absent on gaze to right; caloric response UWR 100 percent; FI, 0.6. OKN not recorded.

Case 14

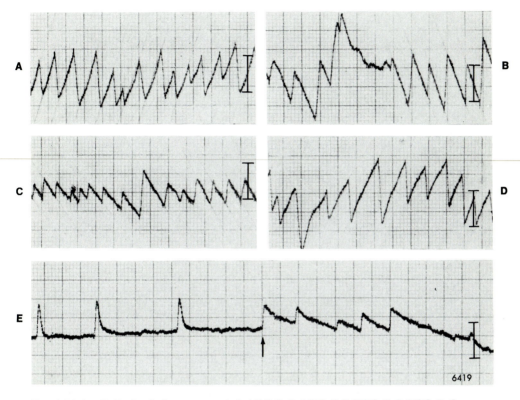

Fig. 11-14. A to D, Peak caloric responses: **A,** L 44° C; **B,** R 44° C; **C,** L 30° C; **D,** R 30° C; **E,** Gaze record, eyes deviated to right; eyes open to left of arrow and closed to right of arrow. All tracings, bitemporal leads. Saccade and tracking tests normal; nystagmus absent on gaze to left; caloric response UWL 6 percent; FI, 0.1. OKN not recorded.

Case 15

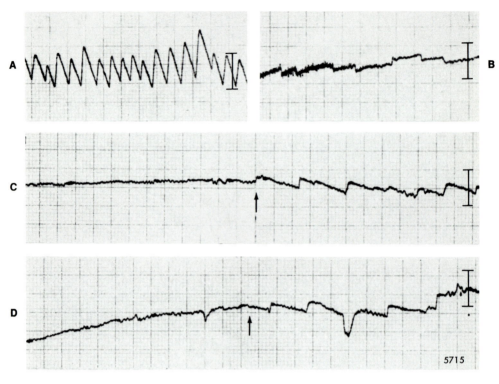

Fig. 11-15. A and **B,** Peak air caloric responses: **A,** L 24° C; **B,** R 24° C. **C,** Gaze center, eyes open to left of arrow and closed to right of arrow. **D,** Gaze to left, eyes open to left of arrow and closed to right of arrow. All tracings, bitemporal leads. Saccade and tracking tests normal; consistent right-beating positional nystagmus, slow component velocity 4°/sec in all positions with eyes closed; caloric response UWR 56 percent; FI, 0.4. OKN not recorded.

Case 16

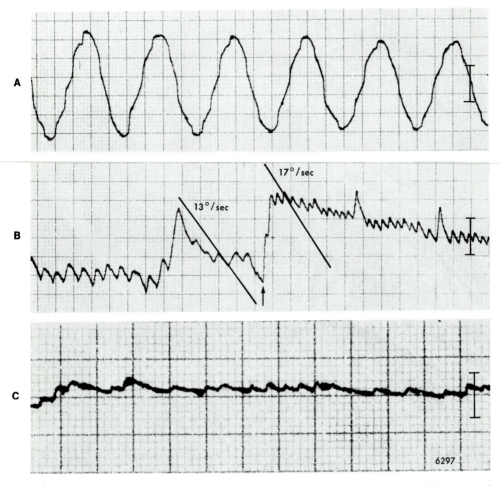

Fig. 11-16. A, Tracking test, bitemporal leads. **B,** Air caloric test, L 24° C, immediately following peak response; eyes closed to left of arrow and open to right of arrow, bitemporal leads. **C,** Gaze upward, eyes open, vertical leads. Saccade test normal; positional nystagmus absent with eyes closed; bithermal air caloric, UWR 2 percent. OKN not recorded

Case 17

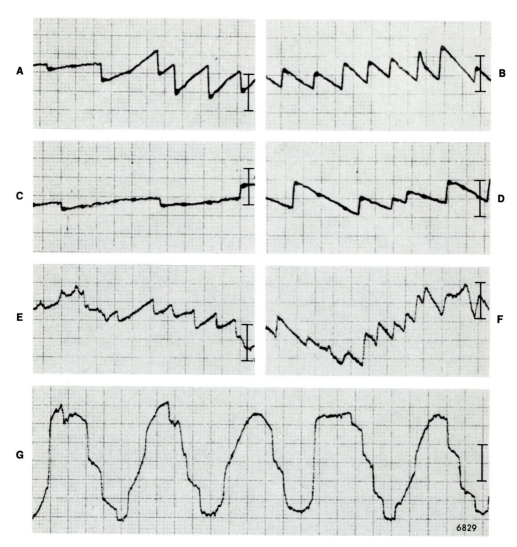

Fig. 11-17. A to D, Optokinetic responses; stripe velocities: A, 20°/sec to the right; B, 20°/sec to the left; C, 60°/sec to the right; D, 60°/sec to the left. E, Positional test, RL position, eyes closed. F, Positional test, LL position, eyes closed. G, Tracking test. All tracings, bitemporal leads. Saccade test reasonably normal; gaze nystagmus to left in primary (center gaze) position and on gaze to left; caloric response UWL 10 percent; FI, 0.7.

Case 18

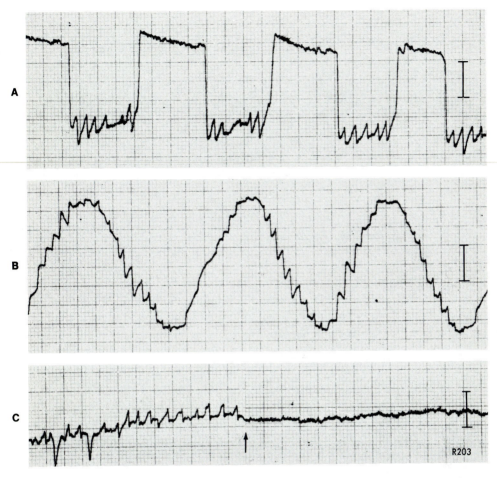

Fig. 11-18. A, Saccade test (horizontal axis). **B,** Tracking test. **C,** Gaze to left, eyes open to left of arrow and closed to right of arrow. All tracings, bitemporal leads. Caloric reponse UWL 5 percent, directional preponderance L40 percent; positional nystagmus absent. OKN not recorded.

Case 19

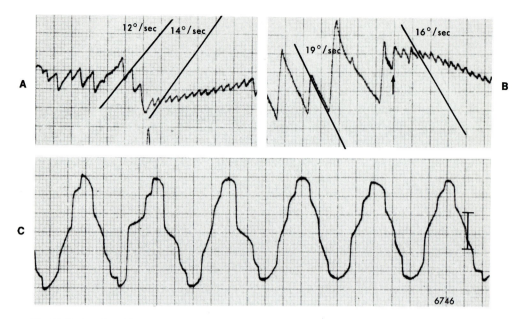

Fig. 11-19. A and **B,** Air caloric responses immediately following peak responses, eyes closed to left of arrow and open to right of arrow: **A,** left ear, 50° C; **B,** right ear, 50° C. **C,** Tracking test. All tracings, bitemporal leads. Saccade test normal; no gaze nystagmus; positional nystagmus absent; caloric response UWR 10 percent. OKN not recorded.

Case 20

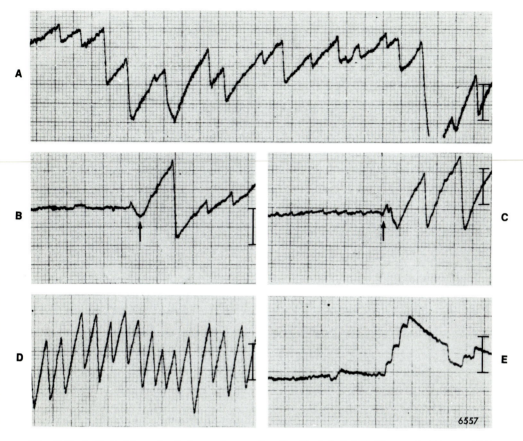

Fig. 11-20. A, Pretest (nonstimulus) record, eyes closed, patient alerted. **B,** Gaze center, eyes open to left of arrow and closed to right of arrow. **C,** Gaze to left, eyes open to left of arrow and closed to right of arrow. **D,** Air caloric, peak response, left ear, 50° C, eyes closed. **E,** Air caloric, peak response, right ear, 50° C, eyes closed. All tracings, bitemporal leads. Saccade and tracking tests normal; left-beating nystagmus with eyes closed unaffected by positioning; OKN normal; caloric response UWR 43 percent; FI, 0.2.

Case 21

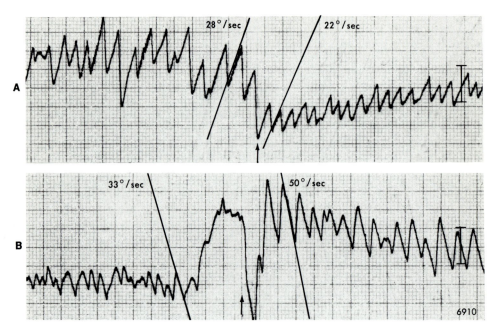

Fig. 11-21. A and **B,** Air caloric responses immediately following peak responses, eyes closed to left of arrow and open to right of arrow: **A,** left ear, 50° C; **B,** right ear, 50° C. Both tracings, bitemporal leads. Saccade, tracking, OKN, and positional tests normal; gaze nystagmus absent; caloric response UWL 4 percent; FI, **(A)** 0.8 and **(B)** 1.5.

Case 22

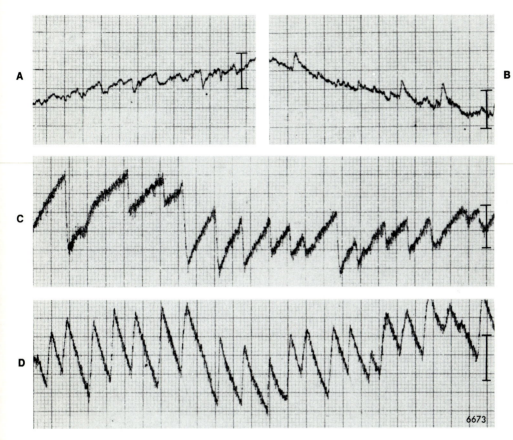

Fig. 11-22. A, Gaze to left, eyes open. **B,** Gaze to right, eyes open. **C,** Positional test, right lateral position, eyes closed. **D,** Positional test, left lateral position, eyes closed. All tracings, bitemporal leads. Saccade and tracking tests and OKN normal; caloric response UWR 12 percent; FI, 0.3.

Case 23

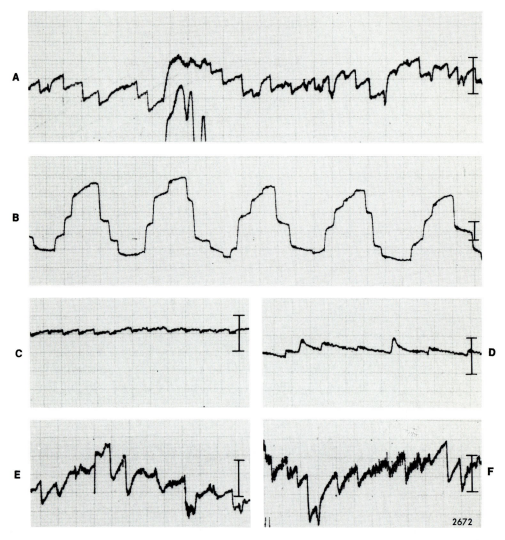

Fig. 11-23. A, Pretest (nonstimulus) record, eyes closed, patient alerted. **B,** Tracking test. **C,** Gaze to left, eyes open. **D,** Gaze to right, eyes open. **E** and **F,** Peak caloric responses: **E,** L 0° C; **F,** R 0° C. All tracings, bitemporal leads. Saccade test normal; OKN pattern is a fairly flat response to increasing stimulus velocity; positional test (eyes closed) shows left-beating nystagmus with slow component velocity of about 6°/sec in all positions.

Case 24

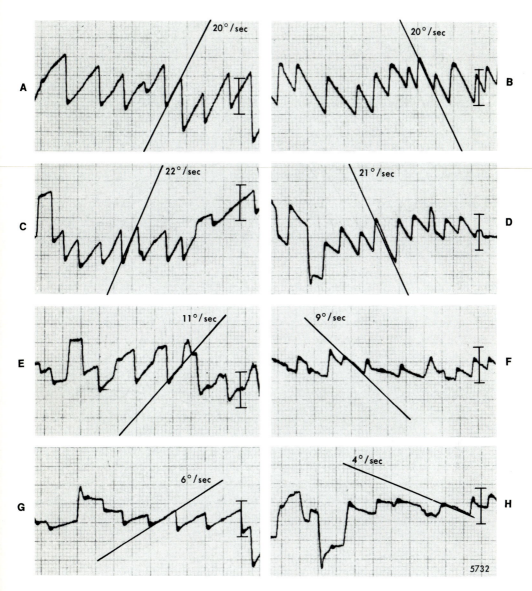

Fig. 11-24. A to **H**, Optokinetic responses; stripe velocities: **A,** 20°/sec to the right; **B,** 20°/sec to the left; **C,** 60°/sec to the right; **D,** 60°/sec to the left; **E,** 80°/sec to the right; **F,** 80°/sec to the left; **G,** 120°/sec to the right; **H,** 120°/sec to the left. Saccade and tracking tests normal; gaze and positional nystagmus absent; caloric response UWR 8 percent; FI, 0.6.

ANSWERS TO QUESTIONS

Case 1

The patient is probably nervous. With eyes open, steady gaze is mantained. The "blips" in the tracing are eye blinks. With eyes closed, square wave jerks are superimposed on right-beating nystagmus, with a slow-phase eye speed of approximately 5°/sec. These tracings reveal no definite evidence of abnormality.

Case 2

a. The patient has down-beating nystagmus.

b. Yes. He probably has a lower medullary or medullocervical lesion.

Case 3

a. The patient has bilateral gaze nystagmus, saccadic pursuit, and failure of fixation suppression.

b. Yes. These findings indicate a CNS (probably brainstem) lesion. The fact that the patient has multiple abnormalities is added assurance that he has such a lesion.

c. Careful inquiry must be made into the patient's history of drug intake. Certain drugs, notably barbiturates and phenytoin (Dilantin), can produce all of the abnormalities seen here. In addition, eye blinks and excessive gaze deviation must be ruled out as the cause of his gaze nystagmus; and lack of alertness must be ruled out as the cause of lack of caloric nystagmus with eyes closed.

Case 4

a. The gaze record contains much artifact, but it shows clear right-beating nystagmus on gaze to the right and probable left-beating nystagmus on gaze to the left. The tracking record shows saccadic pursuit.

b. This patient's medications are probably responsible for her abnormal eye movements. She is epileptic and is in fact taking large doses of Dilantin to control her seizures. She did not discontinue her medication before the test and would have been ill advised to do so.

Case 5

a. The patient has direction-changing positional nystagmus, left beating in the RL position and right beating in the LL position. It is present while the eyes are open but greatly increased in intensity when they are closed. It is strong enough to be clearly abnormal.

b. Yes. This nystagmus is present not only with eyes closed but also with eyes open; it is ageotropic. It is almost certainly caused by CNS disease, provided alcohol was not taken within the preceding 6 to 24 hours.

Case 6

a. The patient has direction-changing positional nystagmus with eyes open only.

b. Yes. This nystagmus is caused by a CNS lesion.

Case 7

The tracing shows congenital nystagmus. The referring physician probably saw congenital nystagmus superimposed on a response of the benign paroxysmal type during the Hallpike maneuver.

Case 8

a. The patient has a severe (79 percent) unilateral weakness on the right.

$$\text{Unilateral weakness} = \frac{(3 + 5) - (32 + 36)}{76} \times 100 = 79 \text{ percent on the right}$$

$$\text{Directional preponderance} = \frac{(3 + 36) - (32 + 5)}{76} \times 100 = 3 \text{ percent to the right}$$

b. One must be certain that the irrigations of the right ear were adequate and that the patient was alert during those irrigations.

Case 9

The patient has a 46 percent unilateral weakness on the right and a 51 percent directional preponderance to the left.

$$\text{Unilateral weakness} = \frac{(4 + 18) - (44 + 16)}{82} \times 100 = 46 \text{ percent on the right}$$

$$\text{Directional preponderance} = \frac{(4 + 16) - (44 + 18)}{82} \times 100 = 51 \text{ percent to the left}$$

Case 10

The patient has a severe, and perhaps total, bilateral loss of vestibular function. The nystagmus seen in these tracings is nothing more than preexisting nystagmus. It did not appear during the gaze or positional tests because the patient was not sufficiently alerted. However, as soon as an irrigation was started, it immediately appeared because the patient was alerted by the blast of water in her ear (Fig. 11-25).

Case 11

All of the responses are beating in the wrong direction. The most likely explanation is that the electrode polarity is reversed.

Case 12

Yes. The FI is $36/26 = 1.4$, which is well outside the normal range of variation.

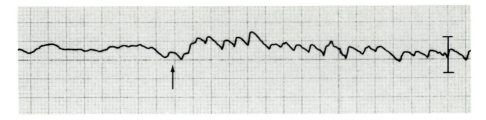

Fig. 11-25. Appearance of nystagmus at the beginning of the warm temperature irrigation of right ear (arrow).

NOTE: In Cases 13 through 24, the final diagnosis of each patient is given. It is important to understand that the diagnosis was always made through evaluation of many sources (history, medical examination, radiography, etc.), and never by the ENG examination alone.

Case 13

SUGGESTED INTERPRETATION: Abnormal. The patient has complete caloric loss on the right and nystagmus on gaze to the left with eyes open only. These results are consistent with peripheral vestibular lesion on the right. The nystagmus on gaze to the left, with eyes open only, suggests CNS damage as well.

DIAGNOSIS: Acoustic neuroma on the right.

Case 14

SUGGESTED INTERPRETATION: Normal, apart from right-beating nystagmus on gaze to right with eyes closed only, which is a rather "soft" finding.

DIAGNOSIS: Acoustic neuroma on the left.

COMMENT: Acoustic neuroma produces ipsilateral caloric reduction in more than 80 percent of the cases. If it is large, it also produces a variety of abnormalities (brainstem or cerebellar) on the ENG recording. The ENG abnormalities are marginal or even absent in rare cases such as this one.

Case 15

SUGGESTED INTERPRETATION: Abnormal. The patient has considerable caloric reduction on the right and right-beating positional (or spontaneous) nystagmus with eyes closed only. These results are consistent with a peripheral vestibular lesion on the right.

DIAGNOSIS: Meniere's disease on the right.

Case 16

SUGGESTED INTERPRETATION: Abnormal. The patient has vertical nystagmus on upward gaze, defective pursuit, and failure of fixation suppression of caloric nystagmus. These results are consistent with brainstem/cerebellar localization.

DIAGNOSIS: Pontomedullary infarct.

Case 17

SUGGESTED INTERPRETATION: Clearly abnormal. The optokinetic nystagmus pattern indicates brainstem disorder. The patient also has gaze nystagmus, defective smooth pursuit, and direction-changing positional nystagmus. These results are consistent with CNS localization of disorder.

DIAGNOSIS: Multiple sclerosis.

Case 18

SUGGESTED INTERPRETATION: Abnormal tracings caused by congenital nystagmus. Vestibular function is probably normal.

DIAGNOSIS: Congenital nystagmus.

Case 19

SUGGESTED INTERPRETATION: Abnormal. The patient has defective pursuit capacity and failure of fixation suppression of caloric nystagmus. This result is consistent with brainstem/cerebellar localization.

DIAGNOSIS: Metastatic carcinoma (breast) in the right cerebellar hemisphere.

Case 20

SUGGESTED INTERPRETATION: Abnormal. The patient has spontaneous nystagmus beating to the left with eyes closed and reduced caloric response on the right. These results are consistent with a peripheral vestibular lesion on the right.

DIAGNOSIS: Vestibular neuronitis on the right.

Case 21

SUGGESTED INTERPRETATION: Abnormal. An isolated but clear abnormality is failure of fixation suppression of caloric nystagmus. This result is consistent with brainstem/cerebellar localization.

DIAGNOSIS: This patient had a recent severe closed head injury.

Case 22

SUGGESTED INTERPRETATION: Abnormal. The patient has bilateral gaze nystagmus and direction-changing positional nystagmus of ageotropic type. This result is consistent with brainstem localization.

DIAGNOSIS: Transient ischemic attack of the brainstem.

Case 23

SUGGESTED INTERPRETATION: Abnormal. The patient has marked bilateral caloric weakness, probably total on the left. The patient also has clear features of CNS disorder: bilateral gaze nystagmus and saccadic pursuit. This result is consistent with brainstem disorder.

DIAGNOSIS: Bilateral temporal bone fracture and severe brainstem concussion.

COMMENT: Bilateral caloric weakness is more common in peripheral than in CNS disease but is not confined to such localization.

Case 24

SUGGESTED INTERPRETATION: Abnormal. The patient has declining response to increasing stimulus speed in the optokinetic test, which indicates brainstem localization of lesion.

DIAGNOSIS: Syringobulbia (medullary cavitation).

COMMENT: The optokinetic pattern improved significantly after surgical treatment.

Index

225